Development Practitioners and Social Process

Development Practitioners and Social Process

Artists of the Invisible

Allan Kaplan

Pluto Press

LONDON • STERLING, VIRGINIA

First published 2002 by Pluto Press
345 Archway Road, London N6 5AA

www.plutobooks.com

Copyright © Allan Kaplan 2002

The right of Allan Kaplan to be identified as the author of this work has
been asserted by him in accordance with the Copyright, Designs and
Patents Act 1988.

British Library Cataloguing in Publication Data
A catalogue record for this book is available from
the British Library

ISBN 978 0 7453 1018 3

Library of Congress Cataloging in Publication Data
Kaplan, Allan.
  Development practitioners and social process : artists of the
invisible / Allan Kaplan.
    p. cm.
  ISBN 0–7453–1019–2 (hbk.) — ISBN 0–7453–1018–4 (pbk.)
  1. Community development—Philosophy. 2. Social change. 3.
Organizational change. 4. Community development personnel—Training
of. I. Title.
  HN49.C6 K376 2002
  303.4—dc21

2001006015

Reprints: 10   9   8   7   6   5   4   3   2   1   0

Designed and produced for Pluto Press by
Chase Publishing Services, Fortescue, Sidmouth, EX10 9QG
Typeset from disk by Stanford DTP Services, Towcester
Printed on demand in the European Union by CPI Antony Rowe, Eastbourne, England

For Mario
My teacher, this time round;
and Brian O'Connell
who lives the outer with such integrity and intuition.

Op geen oomblik is die mens iets wat vas en bepaald is nie, wat net in een koers kan loop nie; maar 'n moontlikheid van nuwe en nog nooit gewese dinge. Ons moet glo aan die moontlikheid dat nuwe waardes, selfs in ons eie tyd, geskep kan word. Ons moet probeer om die verborge groeiplekke van die geestelike lewe raak te sien.

(At no time is a person something that is fixed and given, something that can move only in one direction; but a potential for the new, for that which has never been before. We must believe in the possibility that new values can be created, even in our own time. We must try to discover the hidden growth places of the spiritual life.)

N.P. Van Wyk Louw

It is insufficient to understand our time in merely political or economic terms. To understand what it means to be human obliges a growing awareness of the deepest designs of the soul.

James Hollis

Imagination is more important than knowledge.

Albert Einstein

Contents

Preface

I work within the development sector, within the realm of civil society. Within the non-governmental, non-profit world; within the world of the social. With people striving to transform a society, people working for social change. With what has also been termed the cultural sphere. This world is made up of other worlds as well: the political, the economic, the technological and the scientific. Yet such distinctions are absurd. These worlds are not separate; they interact and penetrate each other, each aspect affecting every other aspect. The social in turn runs through them all. Nevertheless, economics, politics and technology have come to dominate the social; in terms of values, in terms of guiding principles and in terms of methodologies. Our efforts to reduce the social in this way lie at the heart of the growing conflict and contradiction which characterise the world we are making. Sometimes this is driven home to me in a very visceral way.

A few months ago I was in Amsterdam, and came to that city's central square, Die Dam. Usually thronging with tourists, it was now cordoned off; workers were swarming all over it. The entire surface of the square had been ripped up, work was taking place underground on a new Metro system, and clearly once this had taken place Die Dam was to be re-surfaced and returned to its famous originality. There was little room to manoeuvre, and the crowds were ever present on the periphery. This was a highly complex and technological operation, one which must have been carefully planned down to the last millimetre. I stood and watched for a long time, bemused, transfixed. This is how we think that change in the social sphere must be handled too, I thought; thoroughly controlled, bottom lines worked through, all contingencies thought of and accounted for, every last move planned, and the entire operation predictable and costed. It was this kind of operation which dictated our approach to social change.

The absurdity was particularly striking because I had just come from facilitating a workshop with an international development (or aid) organisation, and was still trying to figure out the meaning of what had gone right and what had gone wrong. The organisation had field offices in many countries of the south, and was facing the need for major organisational change. The previous week I had facilitated a very successful process with the directors of most of the field offices in Africa, and was asked to repeat the process with another group of directors from field offices on a different continent. The original workshop took place in Zimbabwe; the latter one in Holland. The success of the original workshop was well-matched by the intractability of the second. I had tried to repeat a process as one might repeat a building operation, and I should have known: the social will not allow itself to be so manipulated.

They were two entirely different groups. The first already had the makings of a team, they had taken the initiative for the workshop themselves, they had chosen and met with me and together we had tried to understand their situation and what was needed; and we had worked in gracious surroundings, under a wide sky, with time at our disposal. The group in Holland had never come together as a team, and I was a facilitator who had been foisted onto them by top management; they had never met me, never agreed to this process (let alone asked for it), and did not feel that they or the organisation needed it. And we worked in a tiny room within a large and impersonal conference venue, grey skies close overhead. In retrospect, I could see the kind of process which might have worked, and taken us through that stuck place; but I had fallen into a trap, been asked to repeat a process so I had repeated it. As if the participants were so many bricks. As if differences between the groups were not real; as if the groups had no independent lives of their own. As if there was no inherent development trajectory, or story, with which I should have worked, unique to each of the groups, despite that they came out of the same organisation.

Staring at the workers on Die Dam, I regretted my stupidity; I had looked the wrong way, been blind to that which was really moving within the group. But I realised too that the trap that I had fallen into was not at all unique; on the contrary, the lack of appreciation for the uniqueness of the social realm has caught us all in a similar trap. Because we understand so little of the uniqueness of the social realm, so little of how to work developmentally within it, we are creating a world of increasing stress and decreasing sustainability.

The problem lies very deep, in our fundamental approach to the world. The twentieth century has witnessed the rise to dominance of a particular way of thinking – that which concerns itself with the control and manipulation of matter. Because we have achieved so much success in our use of the material world which lies outside of ourselves, the way of thinking which supports such usage has come to be taken as the legitimate way of approaching the world. It has come to be taken as given. Yet simply because a particular way works with respect to certain phenomena does not mean that it is universal; it does not mean that all phenomena should be regarded in the same way.

The task which this book undertakes – the advancing of a particular sensibility and practice with respect to social development – lies not within the realm of the material, but within the realm of the social. Here, the kind of thinking which sets out to reduce complexity to simplified component parts in order to control those parts has not proved particularly beneficial, or even reasonable. Yet by and large we persist in attempting a technical, reductionist approach to resolving social situations. Whether we are development practitioners operating within the 'aid industry', managers and leaders of government departments or service bureaucracies or academic institutions, consultants to service, commercial or advocacy agencies, or simply participants in any of these endeavours, we are influenced by the dominant paradigm of our age.

Science itself is moving on. But social and organisational change is still governed by management paradigms taken from big business, management paradigms which themselves derive from the classical sciences, and from economics. The very heart of economics is reductive; if you cannot measure it, it does not exist. Thus social facts too must be reducible to numbers, and controllable. Poverty is simply a technical issue which can be fixed technically; the same is true for marginalisation, alienation, ecological disintegration, unparalleled levels of conflict and pain. While there is growing recognition that our planet is being grossly mismanaged towards impending bankruptcy, there is little appreciation for an alternative approach – one which accepts the specificity of the social on its own terms, and recognises, too, the primary role that social change has to play.

The economic-technological paradigm, incorporated into rational management, attempts to analyse indeterminate and unbounded social situations and reduce them to their simplest component parts – in order to render them determinate and bounded. Then inputs are planned which will target such parts to achieve a predictable and expected outcome. Change is seen as a one-off process which must be controlled and manipulated. We must have the end firmly in mind, try to minimise disturbances from other events which may impinge, and focus on the envisaged product – a new way of working, a new structure, a new policy or practice or strategy.

The philosopher of science Henri Bortoft has noted that 'the fact that modern physics is true – which it certainly is – does not mean that it is fundamental. Hence it cannot be a foundation upon which everything else, human beings included, depends.'[1] To paraphrase the reflective practitioner Donald Schon, the realm of the social is not concerned with solving clear and bounded problems by applying theories and techniques derived from scientific knowledge; rather, we tend not to be presented with problems at all but with messy, indeterminate situations.[2] The construction of a dam, for example, can only be regarded as a material problem to be technically solved if the contradictory social, developmental, economic and ecological dimensions of the situation are ignored; as soon as these are recognised, the indeterminate and ambiguous nature of the issue becomes apparent.[3] Or *should* become apparent, were we not so steeped within a reductive paradigm. Take this poignant sentence from a story of community development undertaken deep in the Kalahari Desert of southern Africa: 'when boreholes sunk deep into the earth had solved the immediate water problem, D'Kar had become so overpopulated and overgrazed that the desert virtually reclaimed the land during the dry season.'[4] Similarly, the conflicting needs and trajectories of different groupings within any social situation, the tenacious distortions of power, the conflicts and contradictions embedded in social life are not amenable to simplistic and linear resolution.

Processes of change are characterised by innumerable extraneous variables. Change sometimes creeps up slowly and insidiously; at other times it falls like a sudden hailstorm from out of a blue sky. Sometimes change impinges on us from radical shifts which occur within our environment; sometimes it is the direct (but opposite or shadow) result of the very things we ourselves have set in motion – our plans have the opposite effect from that intended, and our logic, planning and predictive ability is laid bare. We know, from our own personal development, that our lives are fraught with such contradiction and uncertainty. Our lives are not only under our own control, but are also the ambiguous result of an unfolding pattern which we often can only barely discern the guiding 'causes' of in retrospect.

Vaclav Havel noted, in an address to the World Economic Forum some years ago, that 'What is needed is something larger [than the scientific method]. Man's attitude in the world must be radically changed. We have to abandon the arrogant belief that the world is merely a puzzle to be solved, a machine with instructions for use waiting to be discovered ....'[5] This book is concerned with that 'something larger', with our very attitude and approach to the world – applied to social and development interventions.

There is no doubt that, unless leaders of, and general participants in, social groupings begin to understand and work intelligently with the dynamics which underlie their collective endeavours, the necessity of working in such collectives will prove our downfall, rather than the opportunity which it presents.

This book has been written out of an experience of organisational consultancy, but it is not intended only, or even primarily, for consultants. Whoever you are, and whatever you do, you are responsible for the social reality which surrounds you. You are both leader and follower, and always, inevitably, social practitioner. This is an attempt to explore a new way of dealing with social phenomena; an alternative way of working in the world, for those who wish to have some intentional influence on their surroundings.

The book is organised around four movements. First, learning to observe in a new way; a way which responds to the uniqueness of the social. Second, gaining insight into the patterns which provide social organisms with coherence and continuity. Third, applying this ability to observe, coupled with understanding and insight, to the social organism's actual processes of change, such that we enhance our ability to intervene in such change. Finally, exploring what all this means for ourselves as social practitioners, as members of social groupings, and as developing beings in our own right. Interspersed between most chapters will be found exercises and practices designed to help develop the faculties outlined in the body of the text.

Acknowledgements

There are many I must thank, as influential others. There are a few without whom the book would never have had the courage or inspiration to emerge. I acknowledge by name here just these few:

Mario van Boeschoten, holder of the invisible flame which kindled my heart.

Keith Struthers, for living and working alongside me for so long, on so many groundbreaking projects, for essential conversation, and for introducing me to many of the works which underpin this text.

Sue Soal, friend and co-conspirator on the road.

David Harding, though unbeknown to him, somehow, the keeper of a certain faith.

Sandy Lazarus, John Wilson, Marie Corcoran-Tindill, for reading and encouraging and honing. Siobhain Pothier, for tirelessly editing and improving. Valda West, for the painstaking work of arranging and organising the manuscript.

All of my closest colleagues at the Community Development Resource Association: James Taylor, Nomvula Dlamini, Poppi Huna, Doug Reeler, Peta Wolpe – for the comradeship and for the work which we have created together; also for recognising the value of articulating the practice and so enabling this manuscript to be written.

Sue Davidoff, partner, for unswerving trust and uncompromising penetration, a relentless integrity, and a real ability to discern, for bringing all this to bear on her reading of the manuscript and on the partnership itself.

Social Process and the Practitioner: A Synopsis

Underlying reality there is a world of archetypes.

Richard Wilhelm
(Commentary on the I Ching)

A new way of working with the social is premised on an appreciation for social process, and an ability to work with this underlying and invisible movement. Every living being is in process, which is simply the flow, the stream, of its life journey. Such processes are both archetypal – sharing commonality of pattern with all beings, such as gestation, birth, death and resurrection – as well as unique to the particular being. Individuals and social organisms (groups, organisations and communities), endowed with the gift of (self) consciousness, have the possibility of becoming aware of their own processes, and thus become responsible for their own evolution, rather than merely subjected to that evolution.

Process is dynamic; a river of rhythm and form. It is a pulsing movement, both progression and oscillation, a spiral flow. The process is the whole within which the individual moments occur. It both underlies and emerges out of the parts, and is invisible. More than simply what is directly seen, it is what is sensed, experienced, understood, intuited from what is seen. To apprehend process, we have to move into a different state of being – one which is simultaneously inside and outside, participant and observer, analyst and artist. Such a state of being lies beyond the realm of logic, beyond the reach of analysis, beyond the constraints of intellect. Such capacity entails the development of new faculties and thinking.

Social organisms, being self-conscious and therefore responsible for their own process, can hit impediments with respect to these processes. The flow and movement can become blocked and entangled. It can become arrhythmical, confusing. Or it can become so harmonious as to induce sleep, thus reducing consciousness and mindfulness. It may become so alluring that we attempt to capture it once and for all, in structures and procedures and rules and regulations, which also may induce sleep and reduce mindfulness. It can be defended literally unto death. It can curtail freedom and creativity, rather than promote it. It can lose touch with its changing context – with the wider processes within which it is embedded.

When the social organism experiences a problem, it can be a call for assistance. Such assistance can come in many guises, more or less useful. As social practitioners – whether consultant, leader or constructive participant – we are there to work with the organism's process. As such, we have to learn to read and recognise the underlying patterns, and help unblock or adjust, so that the ongoing process of development may unfold once more. To help unfold what is enfolded within; to enable to emerge that which has become submerged. To allow the organism's path to reveal itself once more, so that it does not turn outwards in its pain, projecting its own weakness onto others; but regains responsibility for itself, achieving openness and tolerance through becoming mindful of its own process (and thereby also of the processes of others).

During any such intervention, which will itself unfold over a period of time, there are many interweaving processes to be aware of, but three are primary: the organism's process, the practitioner's own process of unfolding, and the interaction between these – the intervention – which is itself a process. The practitioner is responsible for maintaining awareness and centredness in all.

We know that there is an ongoing oscillation between order and chaos, fundamental to the very nature of process. That to take on the new we first have to let go of the old. To lose what we have found. To discover, to become aware of the pattern we must often allow the process to dissolve into chaos, so that new order, a new and more appropriate pattern, may emerge. This is true, as well, for these three processes: the client process, the practitioner's process, and the process of the intervention itself. In order to engender development, all these processes will at times lose their coherence, form and rhythm, so as to enable the new to emerge. These periods of chaos are the points of transition itself; without them, the process will not evolve into a new and more healthy pattern.

Inside these points of transition, the coherence of the pattern is lost for a while, so that a new pattern may be found to continue the process. Yet, in the losing, in the chaos, there must still be an outer holding, a wider security within which vulnerability may be engaged with, faced, and moved beyond. It is the practitioner's responsibility to do this holding.

There will be times when the practitioner's own process, and the process of the intervention itself, may become chaotic. When this happens, or when all three processes arrive at this juncture simultaneously, the practitioner is still responsible for holding the outer whole. The practitioner has to be both inside and outside at the same time: inside and outside the organism's process, inside and outside the intervention, inside and outside their own process. There are many different rhythms and forms happening simultaneously, and many different arrhythmical and formless cacophonies sounding, not least the practitioner's own. Centredness is demanded in the midst of such social flux, that the world may still be held, and the thread found once more.

How do we learn such centredness, from which the world may be seen, and intervened into, despite movement, contradiction and confusion? First, we have to learn to see process itself – which is to see the invisible, to appreciate the underlying whole. To see the system as one being, rather than focus on component parts. Then, we have to learn to understand the archetypal patterns which underlie human and social process, and become able to read the uniqueness of individual paths as they manifest within such archetypal patterns. All invisible. Then we have to integrate the discipline of intervention into such social processes, so that it becomes a familiar, rigorous yet flexible practice.

Centredness, self-awareness, means being at home with the notion of emptiness. Not to fill ourselves with opinions and information and expert solutions, but to empty ourselves so that we may allow the social organism's own process to evolve with integrity and rightfulness. The new emerges, it is not created. We can only hope to create suitable conditions from which it may emerge. We can hope to allow and enable, with respect and deference; we cannot impose. The way to deal simultaneously with myriad social processes is to empty oneself. Only then can we attend to the various flows without becoming overwhelmed; this is centredness.

Part I

Observation

Every process in nature, rightly observed, wakens in us a new organ of cognition.

There must certainly be another way altogether, which did not treat of nature divided and in pieces, but presented her as working and alive, striving out of the whole into the parts.

There may be a difference between seeing and seeing, ... for one otherwise risks seeing and yet seeing past a thing.

Johann Wolfgang von Goethe

1
Beyond Reductionism

You can soon become indifferent to song or dance or athletic displays if you resolve the melody into its several notes, and ask yourself of each one in turn, Is it this that I cannot resist? Always remember to go straight for the parts themselves, and by dissecting these achieve your disenchantment.

Marcus Aurelius

Our way of thinking (and consequently, of seeing) takes place within the contextual landscape of our time. We walk within this landscape; its parameters provide guidance, meaning and form. All this takes place largely unconsciously, as part of the 'given' within which we function. It demands tremendous effort of will to step outside these given parameters, to free ourselves sufficiently to see the terrain within which we walk from the outside, to become conscious of the underlying assumptions which we take for granted, and to think (and see) afresh. Consciousness is a hard won and often lonely activity, though uniquely exhilarating.

The landscape, though, forms gradually and unconsciously, until without realising it we are traversing well worn paths as if no other possibilities were open to us. Though social situations are alive and constantly changing, we often see them as inert and static, simply because we have been educated to appreciate the mechanical and the unambiguous. And, appreciating the line of least resistance, it comforts us to reduce and simplify, and avoid the complexities and contradictions, and the open-ended vagueness, of living beings in process. Though nothing seems to stand still long enough to manipulate, we persist in our endeavours because we see the social as material – because we have been successful in dealing with matter. Yet the social is something other.

It is significant that the first section of this work begins with three quotes from J.W. von Goethe. Goethe's genius lay in this: that he 'borrowed his manner of observation from the external world, instead of obtruding his own upon the world'; or, put another way, his view 'always takes its manner of observation, not from the mind of the observer, but from the nature of the thing observed'.[1] There are profound differences between inner and outer phenomena, between dead matter and living process, between the technical and the social. Goethe developed ways of thinking and seeing which lie outside of the dominant positivist paradigm of our age, and which do not reduce, but rather enlarge, our ability to apprehend those phenomena which are imbued with movement and the pulse of life. He laid the foundation for an alternative way of seeing; a way perhaps more appropriate to social phenomena, a way which might be the basic prerequisite for engaging in the art of social practice.

This new way of seeing and intervening cannot simply be appreciated or engaged with unless the bonds which hold us to our current way of seeing are loosened. We cannot just take on the new without creating some space within ourselves by letting go of the old. For this way of seeing is not merely an addition to an already established way – it is an entirely different approach, requiring the cultivation of utterly new faculties. The intention behind this book is not simply to enable you as reader to understand an alternative, but – through adequate engagement – to facilitate the actual development of new approaches and, with practice, new faculties. We have learned to reduce; can we learn to enlarge? We have learned to control; can we learn to respect? We have learned to measure; but not entirely to appreciate. We have learned to plan and predict; but do we know how to enable and allow? We cannot simply struggle against the current status quo from within the paradigms which inform it; we must let go and move beyond.

To let go, we have first to become conscious of where and what we are, of where we stand. It will help to understand the way of thinking to which we have become accustomed, so can we free ourselves; stand outside so that we can think and see in a new way. We begin our explorations, then, with the underlying rationale of classical or Newtonian science.

The successes we have had – for some centuries now – in the control and manipulation of matter, have coincided with the rise of a particular world view, which forms the basis of our dominantly scientific outlook. Currently, with the rise of the new sciences, many previously commonly held (even cherished) assumptions of that scientific outlook are being challenged – a point we return to shortly. But, despite the alternatives, and despite the contrary indications being presented, our way of thinking and seeing itself remains intact – even though it is inadequate to cope with the new discoveries in, and approaches to, science itself. Classical science, concerned as it has been with the usage of matter, has – inevitably, perhaps – given rise to a materialistic mode of thinking. It is this form of thinking which still frames our mental landscape, identified by its sceptical slant, by its reductionist approach, and by its observer – rather than participant – status.

The essence of the materialist mode is informed by a fundamental supposition: that *while we can legitimately apprehend, and comment on, that which is presented directly to our senses, we cannot make other than subjective suppositions about the connections between these phenomena.* In other words, observation of discrete phenomena – the elements which make up our percepts – is capable of validation and objective verification, while the linkages we form between such discrete phenomena – our concepts – are ideas which are internal to ourselves (and our culture) and therefore subjective and 'hypothetical' as information. This is the basic empirical standpoint, as developed initially by the philosophers David Hume and Immanuel Kant.

Hume's philosophy can be summed up in two axioms which he himself described as the alpha and omega of his position. The first runs: all our distinct perceptions are distinct existences; the second: the mind never perceives any real connections between distinct existences.[2] In other words, the only things that can be known about the outside world are in the nature of single, mutually unrelated parts. Whatever may unite these parts into an objective whole can never enter my consciousness; any such unifying factor can only be a self-constructed, hypothetical picture; that is, subjective. We cannot know, we cannot see, the connections between phenomena; we can only develop opinions, and surmise.

Kant distinguished between two possible forms of thinking: the *intellectus archetypus* and the *intellectus ectypus*. By the first he means a form of thinking which 'being – not like ours, discursive – but intuitive, proceeds from the synthetic universal (the intuition of the whole as such) to the particular, that is, from the whole to the parts'.[3] According to Kant, such a reason lies outside human possibilities. In contrast to it, the *intellectus ectypus* peculiar to human beings is restricted to taking in through the senses the single details of the world as such. With these it can certainly construct totalities, but these totalities never have more than a hypothetical character and can claim no reality for themselves. Once again, we cannot see, we cannot know, wholes, we can only know parts. We cannot trust that our concepts – which are the wholes, the meaning, the sense we make of our discrete percepts – ever correspond with reality.

So we are led into the most profound scepticism, without possibility of rebuttal. Matter, in its divided, discrete parts, is all that we have legitimate access to; meaning, sense and any aspiration to understanding the possible underlying connections and formative forces which may give rise to the world we perceive are, quite literally, to be regarded as nothing more than figments of our imagination. Yet it may be asked whether, in so dismissing our own involvement in the world, in so reducing human beings to onlookers rather than active participants, we have not condemned ourselves to a cold, alienated existence: One commentary reads, 'The sciences of inert matter have led us into a country that is not ours ... Man is a stranger in the world he has created.'[4]

Kant's *intellectus ectypus* – a 'discursive' (analytic) rather than 'intuitive' way of thinking – has largely become our mode of being in the world. We believe only in what we can see, at first glance, as it were; in the physical and material. Our tendency is towards analysis, towards the accumulation of information through reduction of phenomena to their component parts. Such reductionism finds great resonance also in our work within social situations, when complex ambiguities are rendered as simple and linear statements, when profound concepts are reduced to boxes and tables and brief one-line, one-word responses, and when the intricacy of sensitive social intervention is contained and packaged as tools and procedures and instruments mechanically applied.

We reduce because doing anything other admits illegitimate assumption into our observations. Yet such reduction removes the connection between the parts from our consideration. We remove the parts from their context, and in so doing lose the sense of their coherence, their integrity, and the underlying impulses which give them life. As Goethe himself puts it, in his monumental work, *Faust*:

> to docket living things past any doubt,
> you cancel first the living spirit out:
> the parts lie in the hollow of your hand,
> you only lack the living link you banned.[5]

And so, though we assume this stance and presume to study life, we focus largely on inert matter. Indeed, so presumptious have we become that many assume life to have come forth from inert matter, and any explanation of it to be reducible once more to inert matter.[6] Such is materialism, the picture of the world we have created; an echoing, lonely, soulless void in which all is reduced to incidental meetings between disparate pieces.

Yet, for all that a materialist science has achieved – which includes, in no small measure, releasing us from the spurious imperatives of fundamentalism and dogma – it is inadequate for the study of living process, and incoherent with respect to its own assumptions.

Classical science has generated a 'thing' view of the world, a mechanistic world view where 'things' act on other 'things' and thus affect those things in ways which are theoretically determinable and predictable. The world is presented as a gigantic clock, where one thing strikes another and causes a third event, and the process is reducible to a set of simple laws which once again theoretically can be described, predicted and controlled.

As regards the study of living process, classical science is being superseded by new approaches which take an alternative route. Led by the new sciences – quantum mechanics, quantum physics, micro-biology – they describe many aspects of the world differently. 'Things' have disappeared; as scientists delved deeper and deeper in their search for basic building blocks they discovered that such blocks, such things, finite and discrete, do not in fact exist. Instead they found that things change their form and properties in relation to each other, as they respond to each other (and to the scientist observing them). This is difficult to grasp, but has irrevocably been demonstrated – the nature of 'substance' is not easily definable, is not one thing; each 'particle' of the world can hold many different, even contradictory properties, depending on their relationship with other 'things'. Thus the world is now seen to consist of 'relation-ships' rather than 'things'. And what we think of as things are actually intermediate states in a constantly changing network of interactions and relationships.[7]

Systems, then – and every living organism is a system – are not reducible and predictable; everything depends on the particular and unique relationships which configure and disappear in an ongoing ebb and flow. We are asked to substitute our notion of predictability for another concept of potential – everything is different and new depending on different interactions, relationships and settings.

The new sciences call for a new way of seeing, or for the legiti-mation of ways which have been denied for far too long. Instead of looking for discrete things, we are asked to develop the ability to look at relationships, at the interactions between component parts. We learn to look, then, not at things, but at the spaces between things, at relationships and interactions and connections. To apprehend the order which moves the whole, beyond the parts.

When we begin to appreciate relationships, the spaces between the parts, then another angle provided by the new sciences becomes available to us. In Newtonian (classical) science, space is regarded as empty, as a void; material reality consists of discrete things which act upon each other across the nothingness of space. But in quantum mechanics it has been proved that 'instantaneous-action-at-a-distance' occurs; in other words, non-local causality is real.[8] There are connections between things which escape us when we think of the spaces between things as empty. But what if space is not a void? We are now developing the understanding that space is filled with fields, invisible mediums of connections, invisible structures, invisible relational webs which influence material things and which provide matter with form. Fields may in fact be more real than matter; it is now thought that discrete particles come into existence, often only temporarily, when fields intersect. These invisible fields, then, are the underlying foundations of reality. They structure space, and it is through this structuring, through such formative forces, that observable reality is made manifest.

Material reality is, then, not the only form of reality. For example, there is a particular type of field called a morphogenic field, which is built up through the accumulated behaviours of species members, and shapes the future behaviour of that species. After some members of the species have learned a behaviour, others will find it easier. The *form* resides in the (morphogenic) field, and it *patterns* behaviour without the need for laborious learning of the skill.[9]

But relationships, connections, invisible fields – all these have been denied existence, and this has patterned our own behaviour so that we cannot hope, cannot presume, to see them. We have begun to speak of them, certainly, within the biological sciences, the psychological and social sciences, and even within the physical sciences. But they have not yet entered our everyday consciousness. Certainly they are not yet sufficiently influential concepts in the world of social intervention. How then can we begin to develop a new way of seeing, one which will allow us to *actually see* – and not simply, and still slightly sceptically, *refer* to – the invisible whole within which the parts are enfolded?

2
Emergence

There is a kind of seeing which is also a kind of thinking ... : the seeing of connections.

Ray Monk (on Wittgenstein)

We said earlier that Newtonian (classical) science, and the world view upon which it is premised, is not only being superseded by new developments but is also incoherent in terms of its own assumptions. An examination of these assumptions reveals that classical science itself, while it denies the possibility of thinking and seeing in a holistic, intuitive way, uses precisely this mode in its own progression.

We have to differentiate between sense impressions – percepts – which are regarded as legitimate and objectively, commonly verifiable, and our thinking faculty – the ability to develop concepts concerning these sense impressions – which is regarded as subjective and personal.

Let us suppose that there sits, watching a game of some kind, some lover of the game and with him, also watching, a tiny child. The grown-up sees every occurrence against a vast, invisible, complicated mental background of all that he knows about the game. If we try to 'unthink' or 'de-think' all that the adult knows, we shall get to something like the sort of picture which is in the consciousness of the small child – mere movements, mere sense impressions, individual and discrete, making no sense, having no meaning whatsoever. Like a person who has never learned to read looking at a page of Shakespeare.

Such immediate sense experience is sheer multiplicity; a medley of impressions; mere juxtaposition in space and succession in time; blobs of smell, of colour and noise, each item standing in isolation; merely particulars. As such, these sense-particulars are entirely meaningless, and disclose nothing of their nature. They are enigmatical, unintelligible entities. And so long as we depend passively, in this way, on what our senses bring us, we are in the dark. Things seem as if they were shot at us from a gun out of the unknown. We are living in a world without values and without meaning.[1]

If we did in fact limit ourselves in this way, our world would be forever unintelligible. It is our thinking, our ability to make connections, to form concepts, which enables us to make sense of our world. Things, discrete particulars, simply mediated to us by our senses, tell us nothing about themselves. They are enigmatic, obscure, lifeless, unyielding. When, however, they are mediated by our thinking, they come alive, and begin to declare themselves and express their meaning. Without our particular mode of consciousness, without our thinking faculty, we know as little about the world as an animal does. We remain surrounded by the impenetrable outside of objects. But standing in our thoughts, we may enter into the inside of the world. We become able to understand, to move behind the individual percepts; behind the scenes, as it were.

Where we rely only on our senses, the world is a cold, isolated place, consisting of single, isolated objects which bear no relation to each other.

> But as it appears to thinking, reality seems to consist of objects that smile on each other; that stretch out helping hands to each other; each communicating gladly with all the rest. Every thought helpfully relates itself to others. We cannot conceive of gold without conceiving also of buttercups and wedding-rings and hair and goodness. We are traveling about the universe upon a magic-concept at lightning speed. Here we find ourselves in a world of associativeness and affinity. Things automatically group themselves. Towards the concept 'organism' rush other such concepts as 'growth' and 'evolution'. We have emerged out of a nightmare multiplicity of single particulars into a world where, of their own accord, things are grouping themselves. We stand now ... amid wholes ...[2]

Using our sense-organs alone, we can see only nature's shop window. But using our thinking we can move behind, go into the workshop itself, as it were, and see where nature is ceaselessly producing the items which finally end up as outer, discrete objects in the shop window. Thus through our thinking do we make sense of the world. The materialist outlook would hold that such sense is always provisional, always hypothetical, always a construct. We cannot know that what we think we see is what we do in fact see. And indeed, our thinking can err; it may seem subjective because it appears in the heads of individual people; we can think more or less effectively, more or less correctly. But it is not therefore legitimate to consign thinking to the realm of the subjective, to mere personal opinion. Our thinking reveals the underlying connections of the world; it is through our thinking that we are able to penetrate beyond particulars into the wholes, the concepts which give the particulars their meaning.

We are not separate from nature, but part of nature; nature permeates everything, including the human mind and imagination. Nature's truth does not exist as something independent and objective, but is revealed in the very act of human cognition. We do not simply impose ourselves on nature, but allow nature to manifest its own order through our participation and involvement, for – following Goethe – *we are the organ of nature's self-revelation*. As our understanding grows, so does that which we are trying to understand – the world itself is evolving through our growing understanding.

Therefore, while we can think more or less correctly, this does not mean that we cannot think, that we cannot see, in ways which reveal the underlying reality of the world. It does not mean that we cannot see the invisible, the connecting fields and relationships. But it does mean that in order to do so, we have to develop our thinking to the point where it can see invisible wholes directly and correctly. It means that we must strive to clarify our thinking so that the world is able to reveal itself to, and through, us. Goethe held that while we must trust our thinking and seeing, we must at the same time develop these faculties so that they become trustworthy. This seeming contradiction was not a contradiction at all for him. It was simply the process by which the world becomes progressively manifest, and revealed.

Let us look at the same argument from another angle. Science, the philosophy of empiricism, and common sense, hold that we know the world through experience. Clearly this is so. But they interpret experience to mean sensory experience alone. Our knowledge of the world comes to us through our senses. Yet while it is true that we could not see the world without our senses, we could also not see it with our senses alone. Knowing even the simplest fact goes beyond the purely sensory.

Take the well-known ambiguous figures used by gestalt psychologists, such as the old woman/young woman, or the duck/rabbit (see Figures 2.1 and 2.2).

Figure 2.1   Old/young

In these cases two different objects can be seen alternately, and yet the sensory experience is the same in both cases. What changes when we see the images in different ways is not anything in the image itself, but the way we organise – conceptually – what we are seeing. The pattern registered on the retina of the eye is the same in both cases; the marks on the paper remain the same. Yet, depending on the way in which we organise what we are seeing, while nothing on the page or the retina changes, everything changes, even into its opposite. And the way in which we organise what we see is not an

Figure 2.2   Duck/rabbit

element in the visual field itself, but the manner in which the elements are appreciated. If we did not organise what we saw in this way, we would be left with an unintelligible and random confusion of individual lines and shapes.

Such non-sensory organisation of perception is in fact the perception of meaning. It is such organisation which makes sense of our world. The individual lines and shapes, discrete, unrelated and meaningless in themselves, take on meaning when they come together to form a whole, or unity, in the way we see them. The marks on the page do not have any meaning at all; the meaning is not on the page, in fact, as it would be if it were a sensory element. In the act of seeing the world, we organise discrete perceptions into cognitive wholes, and we thus see meaning in such unity in a way which is beyond mere sense experience. We see the invisible whole which informs the discrete parts. *We see meaning, not simply sense-impressions.*

There is no fundamental difference between seeing objects and seeing facts. Seeing a horse grazing in a field is simply a more complex instance of seeing meaning than seeing the horse, or the field, on its own would be. We are seeing connections, in all these instances; we are seeing wholes. Yet the whole is not a sensory object – it is that which connects the lines on the page so that they configure into meaning. The whole, the meaning, is not a thing, a sensory percept, but is that which is emergent in the connections and relationships between the things. It is a non-'thing', yet it is not nothing. It is invisible in the individual lines on the page – it does not seem to exist there – yet it is patently visible, for we see it with immediacy.

We both bring the whole to life as well as allow the whole to express itself through us. We move behind the shop window into the workshop, and participate in the creation of meaning even as we perceive the meaning itself.

Such seeming contradictions confound our usual way of understanding our world. So long as we are simply observers, we must indeed approach such contradictions with a profound scepticism. If, however, we conceive of ourselves as participants, as nature's organ of self-perception, then we can begin to embrace the world even as we are embraced by it; we can begin to legitimate an alternative way of seeing.

Classical science cannot progress without making sense, making connections, thinking in terms of relationships. Indeed, discovery in science is always the perception of meaning; it is the coming to a new organisation of what has been observed, rather than a new observation. Yet it holds such thinking to be entirely hypothetical and subjective. Patently, however, we do not see without seeing connections, without organising our perceptions and seeing meaning. The question becomes, how can we begin to refine and clarify our thinking so that we become able to see the invisible wholes which underlie our sense-perceptions? As the whole is non-material, this implies moving beyond the limitations of a merely materialist way of looking at the world, so that we may begin to see the invisible fields and forces and relationships which inform our world.

To begin to do so, and move beyond the contradictions, it helps to appreciate the emergent nature of living systems, not as theoretical constructs, but as directly experienced mysteries. The following chapter will try to illustrate such sensibility, before we return to an exploration of the new faculties required for seeing the invisible.

3
Life's Resources

[T]he power that catches out of chaos, charcoal, water, lime and what not, and fastens them into given form, is properly called 'spirit'; and we shall not diminish, but strengthen our cognition of this creative energy by recognising its presence in lower states of matter than our own.

John Ruskin

The idea in this chapter is not to follow a theoretical argument but rather to open ourselves to the possibility that life is other than merely the chance result of a material combination of physical or chemical elements. It is to help us open ourselves to a sensibility for life itself, rather than to develop an explanation for life and its processes. In our penchant for explanation, we may have distanced ourselves from the breathing world in which we live. We may have forfeited the ability to understand, which is precisely the faculty we need to develop, in order to appreciate life's processes.

[U]nderstanding something is not the same as explaining it, even though these are often confused. Understanding lies in the opposite direction to explaining. The latter takes the form of replacing a thing with something else ... Explanation tends to be reductionistic inasmuch as diverse phenomena are reduced to (explained in terms of) one particular set of phenomena ... Such an explanation evidently takes the form of saying that something is really an instance of another, different thing. Understanding, on the other hand, by seeing something in the context in which it belongs, is the experience of seeing it more fully as itself ... Understanding is holistic whereas explanation is analytical.[1]

We use a mode of seeing which is nothing less than a denial of the wonder and mystery of life, when we look for explanations of life in non-living, inert matter. We may come to believe – and it is so commonly held that it is not recognised as belief or assumption, but merely common sense – that life is created by non-living matter, and can be explained by recourse to the physical (though such explanation still eludes us). We are asked now to develop – or at least

15

to hold as possible – a different form of understanding. As the agriculturalist and scientist E.E. Pfeiffer puts it: 'One should develop an independent sense for life rather than try to explain it as a function or as a result of material causes.'[2]

Let us take water as an opening instance for such exploration. If we analyse water scientifically, we are dealing with dead substance, because dissection brings about destruction of life activities, which are only present in the flowing medium. We can break water down into its component parts – hydrogen and oxygen – but we cannot create water by combining hydrogen and oxygen. For there is a living, formative force which lies behind the phenomenon of water, a force which we cannot see, and cannot replicate, yet which is visible wherever water flows.

We cannot begin to understand the nature of water by analysing it and explaining it in terms of its component parts. Neither could we predict, through knowing the qualities of hydrogen and oxygen, how water, as a composite whole, would behave. Science has tried to explain the unpredictable complexities of behaviours exhibited by systems as compared with the simple predictability of their component parts by referring to the concept of emergence: new, unpredictable qualities emerge when individual elements combine as systems. Certainly new forms of behaviour emerge, and they are not reducible to the system's component parts. But these emergent qualities, these indivisible wholes: are they the chance result of fortuitous combinations, or is there a living world of process and spirit, of form and force, which uses individual components in order to realise itself? Are these components not simply the outer garments, the dead material, which are formed and shaped by life's processes?

When we stand next to a river we may think that we see the river, but in fact we see only one small part. Above where we stand, the river, as whole system, reaches up into the mountains, along any number of tributaries which feed into it ... well, they are 'it', strictly speaking. (Indeed, this is the point – they are both wholes in their own right as well as parts of a greater whole.) And these tributaries have their own tributaries, until they reach right back to the springs and fountains themselves. All of these constitute the living system of a river. But the story does not end even there. The clouds which fall as rain which seeps through the earth until water emerges as a spring are an integral part of the system. The ocean, into which the river flows, gives off much of the water which forms the clouds which return to the river once more. A tree is a living system in its own right, but as it draws water up from the earth tremendous amounts of that water transpire up into the atmosphere, to fall later as rain. The tree is indeed a living system, on one level; but on another it is simply a vessel for the flow of water, an organ integral to – and forming one part of – the living whole which is water.

The stream of sap in trees and plants varies according to the phases of the moon, and according to the moon's position in the zodiac. Water is thus sensitive to cosmic processes in ways which are not accessible to our current ways of seeing and thinking. 'Living' water is able to absorb forces raying in from the cosmos and impart these forces to plants in ways which 'dead' water is simply unable to emulate. This has been demonstrated through many experiments – that water, when it flows freely, is truly full of a vitality imparted to it from the cosmos, whereas water which is constrained and channelled loses this vitality. This is not merely subjective poetic impression; such impressions hold the key of intuitive insight.[3]

Theodor Schwenk, in a lifetime of work with water, has observed that water always tries to form a sphere, a circle.

> Wherever water appears it tends to take on a spherical form. It envelops the whole sphere of the earth, enclosing every object in a thin film. Falling as a drop, water oscillates about the form of a sphere ... A sphere is a totality, a whole, and water will always attempt to form an organic whole by joining what is divided and uniting it in circulation. It is not possible to speak of the beginning or end of a circulatory system; everything is inwardly related and reciprocally connected.[4]

At the same time, water is subject to earthly laws; it follows the pull of gravity, always seeking a lower level – and it is this tension between the life force of its own nature and the earthly laws to which it is subject that creates, for example, the meandering pattern taken by a river as it flows seaward. It is not simply material obstacles which form such patterns, but the combination of these with water's unceasing endeavour to form a sphere. This endeavour lies beyond the material world – it expresses the living, formative force which provides water with life, the *form* or *field* which patterns the behaviour of the sytem.

We should not, then, disregard the feelings which arise in us when looking at a pure mountain stream as mere subjective poetry. We can develop our feeling into a real seeing, a real sensibility, through opening ourselves to the correspondence of this seeing with the actual reality of the water. Thus we can penetrate through the appearance of water to its actual nature. All organic wholes are powered by fields or formative forces which, while invisible, provide the underlying resource which manifests in what we see.

Those of us who have access to mountains and wilderness recognise immediately the difference between 'living' water and 'dead' water. The water flowing from our taps in the cities, or through concrete canals, while it may be clean, is lifeless and dull. On the other hand, water flowing freely in the mountains, bubbling and gurgling from rock to rock, is alive and joyous and full of vitality. The point is that if we approach water with a childlike wonder, with poetic intuition, if we do not try to reduce and explain but rather plunge ourselves into looking without prejudice, then we can see for ourselves the difference between living water and dead water. Chemical analysis will not detect this difference, but our own sensibility needs no such analysis. Reading water in this way, we can begin to understand the importance, for living process, of its free-flowing nature.

There are mysteries on this earth which science does not have the power to explain; for life is not the result of an arbitrary chain of circumstances but an autonomous power. As indicated above, each organism is a self-contained entity, but forms part of a larger whole. And it is the whole, always the whole, which is imbued with life, and with living process. All trees in a forest are one organism, even though they are individual trees. Those that belong to one species, no matter where they grow, are part of one organism. There is a certain bamboo which blooms once every 18 years on the Amazon River. Sprouts of this tree were taken to England, Switzerland, Germany and Java. Even if the seasons were quite different the shoots of this bamboo bloomed exactly the same time as in Brazil, every 18 years. The red-leafed beech grew in Italy, but from there has been planted all over Europe and the United States. At one time the mother trees became diseased in Italy, and died; and all their descendants, whether young or old, started to get diseased and die, wherever they were.

All plants of one species form an organism, all plants together form an organism at a higher level, all life on earth is again part of an organism on a higher level, and each part is an organ. Life forms one indivisible whole, and the energy of the forest or the power which organises life cannot be found in the physical world. It is invisible, yet manifests itself. It lies behind, beneath, and brings forth the world as we know it.

I write this in a cottage in the mountains, and every day when I walk the hills I come across the mounds of termites – hard, round, black mounds about knee-high. Termites have evolved an amazingly complex and efficient social life, with hundreds of thousands of individuals living in close, interdependent association within colonies that are run on absolutely totalitarian lines, with a single queen communicating her instructions across seemingly impenetrable objects. They live entirely, and continuously, underground, in the pitch darkness of their air-conditioned mound, never seeing the light of day except on one occasion each year. This occasion is when the nymphs – specially prepared males and females – are set free in a wedding flight, from which new colonies are formed. In the mountains of the Western Cape province of South Africa in which I write, the preparation of the nymphs begins in January. They are kept prisoners (by the worker termites) for several weeks, for they are not allowed to leave the mound until the climatic conditions are just right for the wedding flight. Then some time in March or April the first winter rains arrive at the Cape, and upon a day when the weather is clearing after a good rain has soaked the ground, when the air is still and the temperature rising, the long-awaited signal is given for the great event of the year to begin, and the top of the mound is pierced in many places to allow the nymphs to emerge into daylight. We do not know who or what decides that the time is auspicious. Nor do we know how the termites, hidden in the blackness of their air-conditioned home, learn what the weather conditions are like outside. But the signal is given, and the activity begins, not simply in one colony, but simultaneously in all the mounds in a given area – mounds which are not connected in any physical or material sense.[5]

And again: homeopathic (medicinal) preparations are made by dissolving particular material substances in liquid, and then diluting the resultant mixture again and again, until eventually no chemical trace can be found, in the liquid, of the original material substance. During the process of ever-increasing dilution the mixture is subjected to succussion, a pulsing movement taking place with a particular rhythm. As the dilution increases, as the material substance disappears, the potency of the preparation becomes stronger and stronger. The greater the dilution, the higher the potency.

Homeopathic preparations thus do not depend on material substance for their effect, but gain in effect as material substance is diminished. When no material or chemical substance is present any longer, the dilution is most potent. Something is at work through this process. As the material substance disappears, its life force – call it spirit – emerges more fully, and it is this life force, devoid of matter, which is more powerful than the material substance or chemical composition with which we started. With homeopathic preparations, then, we are in the presence of pure life force, the pure underlying process which lies behind substance, and which gives form to it. We have released this spirit from its material manifestation.

Wherever we look, there are energies which are unknown to us through regular scientific investigation. *The forces of life organise matter.* Calcium, potassium and phosphates are taken out of the soil by the roots of a plant and carried at the speed of one or two feet *per minute* into the stem and leaves. Our body takes up matter in a similar way and transports it very quickly through all its parts. If we place irradiated sodium chloride under our tongue, it can be traced in *one minute* all through our bodies. Yet our blood needs 2.5 minutes to circulate through our body.[6] It has been discovered that subatomic particles such as electrons are able to instantaneously communicate with each other regardless of the distance separating them. Whether they are ten feet or ten million miles apart somehow each particle always seems to know what the other is doing.[7]

Such instances all point to the fact that the formative forces and processes of life infuse matter, and invisibly form and mould what eventually appears to our senses as thing, as substance or discrete entity. Such examples are not given to prove any particular point of view or theory, but rather to develop a sensibility for the processes of life which, though invisible, infuse every organic whole, every system; which act as the formative force creating the structures and substances which we encounter. We are asked to turn our thinking around. Baron von Herzeele, a philosopher and scientist and follower of Goethe, noted that 'What lives may die, but nothing is created dead' and 'the soil does not produce plants; plants produce soil'.[8] As Pfeiffer puts it:

> Not the chemical substance of which things are made but the imprint of the formative principle is what gives shape and existence ... There is no such thing as completely amorphous matter. Wherever matter exists, it takes form. The formative force

which gives shape to matter is the primary principle antecedent to the aggregation of matter itself. Without it no matter can exist, and without recognition of the existence of the formative principle in matter one cannot understand matter itself.[9]

Overcoming a materialistic view of nature means learning to see phenomena freshly, practising a way of looking that is an active 'thinking into' things – not merely a recording of measure, weight and number followed by explanations that imprison facts in a rigid world of hypotheses and theory. Such a sensibility will not be found where the attempt at explanation reduces living wholes to their component parts. It is not explanation we are after, but understanding, and here context, connection, relationship and invisible field are all-important. Such understanding demands an appreciative rather than a sceptical frame of mind. It is not attainable without it.

Working as a consultant with organisations and social groupings, and as a leading practitioner in my own organisation, has encouraged me to recognise the life processes which underlie such social organisms. Here too are invisible, formative forces which provide the living breath of process manifesting as organisation or community. In these systems, qualities abound which are unpredictable, uncontrollable, and to a large extent invisible; yet they can be understood and seen, so long as we do not simply reduce them to their component parts. Thus, with the right appreciation, such processes can be guided.

Because we are steeped in a materialistic, empirical mode of seeing, we tend to look immediately for that which is most tangible; and we then, if we are wanting to assist or intervene, often approach such tangible parts in piecemeal fashion, as if they were discrete entities which could be remedied, or changed, on their own, perhaps with little effect on the system as a whole. Thus we tend to concentrate on organisational or community structure and procedure; we focus on material resources or tangible skills, we isolate elements of strategy or impact; we engage in quantifiable assessments, and stated needs.

Yet all these are merely the outer, visible, manifestations of invisible fields of force which pattern and form organisational and community life. Of far more importance, and import, are the intangible webs of relationships and connections which subtly lead the social organism to the particular place in which we find it at any given time. The dynamics of power, the complexities of relationship, the quality of leadership, the understanding that the organisation has – or does not have – of its identity and place in the world; the sense of team or conflict or hierarchy, the interplay of dependencies and independencies; the age, the connection with context, the quality of impact and interaction – all these factors form the invisible skein, the network or web of fields and forces which pattern and form the organism in its systemic whole.

These elements act as invisible fields which pervade the social organism and underlie the behaviours it manifests, as well as the material and structural aspects it chooses to adopt. Taken together, they create what is known as the culture of the organisation, which is really nothing less than the soul, that complex essence of the organism's being which mediates between spiritual possibility and material constraint. The group's culture is the expression of its being, and this being exudes a force which permeates the group even as it is formed through the complex interweaving of the fields mentioned above.

To really work effectively with social groupings requires an ability to apprehend directly, not only structures and procedures and humans-as-resources and material conditions, but connections and relationships and formative fields. John Heider, in *The Tao of Leadership*, has this perspective:

> Pay attention to silence. What is happening when nothing is happening in a group? That is the group field. Thirteen people sit in a circle, but it is the climate or the spirit in the centre of the circle, where nothing is happening, that determines the nature of the group field ... People's speech and actions are figural events. They give the group form and content. The silences and empty spaces, on the other hand, reveal the group's essential mood, the context for everything that happens. That is the group field.[10]

When we form social organisms by living and working in them, we call into being living forces beyond our individual ken or control; new worlds, new beings, new forms of life are enabled, greater than our individual selves, unpredictable and immense. New living processes are activated and made manifest. New wholes are created, with their own indelible power. To say this is to assert far more than simply the observation that (organisational) culture is more significant than structure. It is to claim that every social organism – however small, however temporary – is a new instance of life striving to manifest its being, clothing itself in a particular social form. It is to this life, to the whole in its process of growth and becoming, that we are asked to attend.

4
Intuition: Seeing Holistically

To the rationally minded the mental processes of the intuitive appear to work backwards.

Frances Wickes

There are two major modes of human consciousness which are complementary.[1] One may be termed the *analytic* mode, the other the *holistic*. The *analytic* mode is that mode which characterises what we have referred to as Newtonian or classical science; what we have called a materialist, reductionist approach to the world. The *holistic* mode of consciousness is that which I am calling a new and alternative way of seeing; one which is more appropriate to observing life, wholeness, underlying process.

The analytic mode is the mode we have specialised in, the mode to which our educational system – within our technical-scientific culture – is geared almost exclusively. This mode develops in conjunction with our experience of perceiving and manipulating *solid* bodies. The internalisation of our experience of the closed boundaries of such bodies leads to a way of thinking which naturally emphasises distinction and separation. In the world of solid bodies, everything is external to everything else, and it is this separation which contributes to a reductive and analytic frame of mind. This way of thinking is also, consequently, sequential and analytical, proceeding from one element to another in a piecemeal fashion – the principle of mechanical causality is thus a typical way of thinking in this analytical mode of consciousness.

The principles of logic are associated with this way of thinking and seeing; that one thing is always itself and never anything else, that something cannot simultaneously be itself and not itself, and so on. These principles – derived from perceiving solid objects, and from standing outside of the world, as observer rather than participant – are extrapolated and assumed to hold universally.

The emergence of the analytic mode of thinking has been of great significance to the evolution of consciousness. It marks the process of awakening from an unconscious or preconscious unity to a conscious separation and division. It literally places us 'outside', and has thereby not only facilitated practical and material expertise, but has also emancipated us from much of the hierarchy which accompanied the dogmas and fundamentalisms of tradition. But it can no longer, on its own, meet all our needs, for the world with which we are concerned when we are dealing with living organisms in their processes of development and change is not a world of solid objects. We cannot 'see' the life of a termite mound in this way; we cannot in this way 'see' the unified energy of a forest, or mark the difference between water which is living and water which is dead. We cannot 'see' an organisation in this way, or a social group. We can see individual people, a building, equipment, written statements of mission, strategy and organogram, the name of the organisation on its building; but we cannot 'see' the organisation. (*Where* exactly *is* the organisation?) We cannot 'see' connections, evolving relationships, fields of influence, the web spun by habit and routine.

In the normal sense, we can see discrete objects – although already here, as noted in Chapter 2, we are organising our perceptions, forming concepts. But we cannot – in the same way – 'see' the connections, the lines of force and the relationships between them. We can see the thing, but not the meaning of the thing. We have to use our conceptual ability, holistic consciousness, to see the whole, the meaning itself. Music remains incoherent noise if it is heard in fragments; apprehension of the whole renders noise into music. The same can be said of a poem, a joke, a painting, any interaction between people.

The meaning, the *gestalt*, the whole, is not a discrete object in the way that the parts which go to make it up are discrete objects. And it is not simply the sum of discrete parts, or objects. We do not appreciate the whole by adding the parts together; it is observed directly, on its own ground. It is intangible, of a different order, and demands a different approach.

The world with which we are working when we work developmentally and socially is a world of intangibles. It is a world of systems, of relationships, of connections; ambiguous, shifting and changing, developing, interweaving, continually being formed and continually dying and changing into something else. In a word, dynamic. It is not a world of objects but a world of relationships between objects, a world of character and meaning.

The holistic mode of consciousness is complementary to the analytic one. It is systemic thinking. By contrast, this mode is non-linear, simultaneous, intuitive instead of intellectual, and concerned with relationships more than with the discrete elements that are related. It is important to realise that the holistic mode of consciousness is a way of seeing – or, in fact, *organising* our seeing – and as such it can only be experienced in its own terms. That is, it has to be *experienced*, not simply thought about from out of our normal analytic mode – we will move to questions of practice later. We have to appreciate what it means to say a *relationship* can be experienced as something real in itself.

This is not as complex as it seems, and in fact we do it all the time – as we have already seen – *but we do not value it,* or do it consciously. Because we do not value it, we do not cultivate it, and therefore it atrophies, stagnates, and we remain incapacitated and unskilled in a profound sense. It becomes ridiculed as unscientific, unprofessional, unaccountable; we come to believe that such a way of seeing cannot be replicated (that it is nothing more than subjective opinion) or taught; that it is quirky, or arty; that it belongs to the realm of religion (where Kant consigned it), or spirituality (expressed pejoratively). It is none of these things, but a disciplined and cultivated form of what, if we are honest, we do all the time. And in learning to see in this way we change nothing of what is seen, but *everything* changes, because we see that which was invisible to us before.

Intuition is not intangible or mysterious; it is *the simultaneous perception of the whole.* We can learn (or re-learn) to do this. For we must be clear on one thing. The development of the new sciences has shown that we affect what we see by what we are looking for. The particle (separate and discrete, and so small that it has almost no extension) is also seen as a wave (which can extend to infinity, and be entirely penetrated by other entities) *depending on what we are looking for.* The act of observation changes what is observed.

We are participants in our world's unfolding, not simply onlookers. We are part of what we observe, and thus affect both it and ourselves. We will enable to emerge what we look for, and eradicate what we do not look for. If we look only for superficiality, for efficiency, for number, for structure, for the discrete object, then we will create a world which is devoid of the invisible breath of life, of wholeness and meaning. We will reduce our world to a world of inanimate things.

Observation is thus a moral act. We are implicate. We cannot assume that the world is the way it seems; we allow it to emerge as we go, choosing the world we want by how we choose to see it. Denying the value of the holistic mode of perception reduces our world to an empty and inexplicable shell. Embracing the invisible will allow us to work as social artists in a world which is at once mysterious and meaningful; ever present and always becoming; ever alive.

What may this mean for seeing social organisms themselves? As an illustrative instance, let us consider the following question: is a particular organisation, working in the social sphere, sustainable? We tend to address this question by looking at discrete parts – particularly its finances and its future sources of financial supply – analysing those parts and constructing a 'theoretical' position through calculating what the sum of those parts might mean. Yet this sum gives us only the most basic foundation from which to work. An organisation's sustainability depends really on its robustness, its ability to respond intelligently to its changing circumstances. Here a whole host of factors come into play – vision, strategy, leadership, history, staff capacity, internal values and habits. All of these cohere into a particular 'character' which will determine the organisation's sustainability.

Yet the perception of such character cannot be arrived at through reduction to particular factors and then, through linear addition, generalising towards the 'sum' of such factors. Learning to see will enable us to perceive such character directly, for such character is one, and whole, and indivisible. The organisation, too, can be enabled to directly perceive its own character. Does it tend to regard problems as being 'out there' rather than its own responsibility? Or does it regard itself as the creator of its own destiny? Is the organisation introvert or extrovert? Does it lean towards principle or expediency? And so on. Once such character is perceived, work can be done on its component parts, the invisible and visible factors which constitute its character. But the idea is to proceed from the whole to the parts, and not from the parts to the whole.

Ludwig Wittgenstein, whose later work was much influenced by Goethe, talked of this experience of holistic seeing as that understanding which consists just in the fact that we see the connections.[2] Wittgenstein recognised that the idea was not to search for new generalities or theories, but to develop a new *kind* of seeing. Seeing holistically is a higher cognitive function than abstracting what is general; it is the ability to see connections *directly*, and thus experience the whole. And it is the ability to recognise that such wholeness is an intrinsic part of the phenomenon we see, and not added to it by the mind, though the mind enables it to emerge.

If we return to the images presented on pages 12 and 13 and consider someone who is unable to see one or other aspect of these images, but who sees only random black and white blotches, we have a situation in which Wittgenstein would describe the person as *aspect*-blind – as also in those who cannot understand a joke, or appreciate a piece of music or poetry. Wittgenstein asks what is lacking in these cases, and answers: 'It is not absurd to answer: the power of imagination.'[3] For Wittgenstein, we *see* a connection in the same way we see an aspect. For both Wittgenstein and Goethe, the kind of seeing which *sees* connections is *imagination*. 'Imagination is the kind of seeing which is also a kind of understanding (a kind of thinking).'[4] For imagination, seeing and understanding are one. And what we see is not like a physical object; it is that which connects particulars to each other within their context, it is the indivisible whole and meaning, the pattern or field.

Imagination, then, is not to be dismissed, as in: 'merely a figment of the imagination'. The imaginative faculty needs to be enhanced and developed such that it may apprehend what is really there, and enable to emerge what is striving to manifest. And whereas these two kinds of seeing – analytic and holistic – are different, they are not exclusive. We can and do have, and use, both sensory and imaginative seeing together. Imaginative seeing does not replace sensory seeing but is present along with it. William Blake referred to this kind of vision as 'twofold'.

As an illustration, and to recognise that we encounter such twofold seeing all the time and are already familiar with it, consider the activity of reading. When we read a word there are several distinct letters, but the letters are seen as a word only when the discrete elements are seen holistically. The individual words are seen as a sentence when the distinct elements are seen holistically. What is seen, thus holistically, is the meaning of the word or sentence, which is seen as a whole – and this is clearly very different in kind from the individual elements. The physical manifestation is experienced through the senses, whereas clearly the meaning is not; one needs to bring the imaginative faculty into play.

When we see the meaning in this way, we know full well that what we are seeing with our physical sense of sight is not the meaning but the physical marks on the page in front of us. And the physical marks do not disappear when we read the meaning. We can – and must – see both at the same time, else we could not read. When we understand what we are reading, we are seeing something which is very different in kind from the marks on the page. Yet we can only come to the meaning – the whole, the intuited connection – by means of the discrete marks (the physical manifestation) through which it is expressed.

The ability to read is a developed faculty, and certainly makes use of imagination as an enhanced form of thinking, without which we would understand little of a sentence or word or text. But this does not mean that the meaning of what we read is simply a spurious figment of our imagination. It exists as surely as do the marks on the page. And the more we understand, the more we are able to see, the more meaning will emerge. In like manner, we are called upon to develop the capacity to see holistically in order that we may read the living phenomena of our world, and become able to see the formative processes which underlie the development of social organisms.

5
Indications for Practice

The means whereby to identify dead forms is Mathematical Law.
The means whereby to identify living forms is Analogy.

Oswald Spengler

The isolated mountain cottage in which I write is built entirely from unhewn stone dug out of the mountainside itself, dry-packed in the traditional way. It encloses me with nuanced colour, textured patterns and undulating angles, natural, rudimentary, as subtly moving as a work of art. A warm mosaic of earthen tones and tumbled forms. I think of Mr Machachi, the old stonemason whose mastery made this cottage what it is. I remember watching him with awe as he built, slowly and patiently, moving at the pace of the rocks themselves. This was no bricklaying exercise. He would stand and contemplate the wall before him, and then turn to the monumental pile of rocks which we had brought down from out of the higher hills, and from that incomprehensible heap he would choose one rock, and place it in the space which was waiting for just that rock in the unfinished wall. Sometimes he would have to chip a bit here and there, sometimes add a few small stones for balance, and a few handfuls of mud to hold it in place for a while. But always, the rock he chose was the right rock for the next space available. Then he would contemplate the wall again, for with the additional rock the entire configuration of the wall changed, and became new, and he would have to read the situation afresh. How different such work is from the construction of a pre-planned brick wall, and how startlingly different the result. The cottage feels whole, as though it were alive, as though it were organic.

Some few hundred feet above the cottage, before the mountain takes its gigantic strides to the summit, there is a gentle hollow in the hill from which I can look down off the mountainside on to the valleys below, which gradually stretch out into long cultivated plains extending to the horizon. Some distance away a lonely road winds its way towards me from out of the blue horizon, briefly touching the outer reaches of the mountains before turning and moving off around the green hillside once more. There must once have been no road, nothing to break the natural whole formed by an harmonious landscape; but the road does not intrude into the solitude – it does not impose itself, but, snake-like, follows the contours of the hills and valleys. It feels, as it were, slightly built, and a new whole has been formed of the landscape, one which includes people and our various activities. I conjecture, sometimes, what the view would feel like if a four-lane, modern highway were to replace the road, with a straight concrete bridge spanning the central valley, from one side to the other, to avoid the time consuming length demanded by the contours of the landscape. For me, I know, the natural integrity of the scene before me would be destroyed; the whole which would emerge from such a juxtaposition of parts would be hurtful and, I would claim, even unhealthy. A whole would have been broken, forever and irrevocably. Another whole would have been created, yet with internal tensions which would generate entirely new relationships.

I mention these observations because it feels important to emphasise that we are creating wholes all the time, whatever we do and wherever we are. We may pay little attention to it, and focus more on the specific activity at hand than on the way in which it relates to everything around it, but all the time we are creating wholes, creating meaning; and it may help to become aware and concerned about just what the meaning is that we are creating. To look at the same thing from the other side, whenever we do anything, alter anything, we are enabling the world to express itself in new ways; we are enabling new forms of life to emerge, and these all have their effect on us, on the living world and on the social fabric which surrounds and contains us. As Ralph Waldo Emerson puts it:

> The charming landscape which I saw this morning is indubitably made up of some twenty or thirty farms. Miller owns this field, Locke that, and Manning the woodland beyond. But none of them owns the landscape. There is a property in the horizon which no man has but he whose eye can integrate all the parts.[1]

Our world is not yet fully formed. It is in the process of becoming, and we are creating it as we go, not only through how we intervene, but through the way in which we apprehend the meaning which is emerging. If we are to take responsibility, we must be careful, and observant; it becomes a matter of necessity that we pay attention, at all times, not just to particulars but to patterns and connections and configurations, because these constitute the invisible wholes, the meaning, the emerging life and world which is our future. We live within intangible fields and vast formative processes which pattern and shape our very existence.

For those of us whose work lies in the social field, the task becomes urgent and immense. We have to learn to read social process as the stonemaster Machachi reads his walls – with patience, understanding and respect for what is emerging as the process of becoming unfolds. Else we will create social groupings which are inherently constraining, oppressive and distorted. How many of us live and work in such social organisms already?

There are many ways of developing the capacity to see holistically, to see the living whole, to read and understand the meaning which lies invisible yet fully revealed all around us; as Goethe put it, nature is an open secret. All these ways have to do with developing our capacity for observation, reflection and participation. Though we are looking for that which lies beyond the sense particulars, nevertheless we do not see it through abstracting, generalising and intellectualising. We are looking to develop an intimacy which *discloses* the world.

The first thing is simply to be aware of the difference between wholes and parts; to become aware of what we are looking for. Also to be aware that we have the capacity to see what we are looking for. And to appreciate this capacity; rather than to denigrate it as unscientific or subjective or fanciful. We are not looking for the thing, but for the spirit of the thing. We are looking for character, or essence, or meaning. J.D. Salinger relates the following Taoist tale:

Duke Mu of Chin said to Po Lo: 'You are now advanced in years. Is there any member of your family whom I could employ to look for horses in your stead?' Po Lo replied: 'A good horse can be picked out by its general build and appearance. But the superlative horse ... is something evanescent and fleeting, elusive as air. The talents of my sons lie on a lower plane altogether ... I have a friend, however, one Chiu-fang Kao ... who in things appertaining to horses is nowise my inferior. Pray see him'. Duke Mu did so, and subsequently dispatched him on the quest for a steed. Three months later he returned with the news that he had found one ... 'What kind of horse is it?' asked the Duke. 'Oh, it is a dun-coloured mare,' was the reply. However, someone being sent to fetch it, the animal turned out to be a coal-black stallion! Much displeased, the Duke sent for Po Lo. 'That friend of yours ... has made a fine mess of it. Why, he cannot even distinguish a beast's colour or sex! What on earth can he know about horses?' Po Lo heaved a sigh of satisfaction. 'Has he really got so far as that?' he cried. 'Ah, then he is worth ten thousand of me put together. There is no comparison between us. What Kao keeps in view is the spiritual mechanism. In making sure of the essential, he forgets the homely details; intent on the inward qualities, he loses sight of the external ... He looks at the things he ought to look at, and neglects those that need not be looked at. So clever a judge of horses is Kao, that he has it in him to judge something better than horses.' When the horse arrived, it turned out indeed to be a superlative horse.[2]

Although in this instance elevated through practice to a form of mastery, this is what is implied in seeing wholes.

To begin to do this, we must turn the activity of seeing inside out. Normally we take the activity of seeing quite passively; we simply allow the world out there to present itself to us. Instead of doing this, we can begin to see actively by consciously reversing the action of seeing through projecting it outwards towards the phenomenon, rather than merely receiving the impression passively. In this way we can learn to *plunge* ourselves into seeing, as it were, and in so doing enable the phenomenon to reveal itself in all its diversity and in its essential character. This takes time, patience and effort. Yet if we are indeed the organ of nature's self-revelation, then we may approach the matter in the understanding that the world *wants* to reveal itself to us. (And for 'world' we must also read every social situation, every social organism, with which we may be concerned.)

In this way, we can begin to pay attention to the natural and built environment which surrounds us. Look at the coherence and disjunctures, the form and flow and rhythm inherent in the way things fit naturally into their context, the patterns and fields that unite and connect; or at the lack of harmony and rhythm, the interrupted flows and the crude blockages which create different kinds of wholes, if that be the case. If we plunge ourselves into seeing and are not afraid to use our imagination, then the world will begin to reveal itself.

Seeing in this way, whether it be the natural world or the built environment or the social or economic environment, we begin to penetrate through to the realisation that indeed all is relationship, that the world is made up of patterns of relationships, and that what we thought were things are actually relationships between smaller elements, which are themselves relationships between yet smaller elements. And we begin to see how repeating relationships lead to patterns which form the world around us, and give it its sense of stability.

To help with 'feeling our way along', this book must provide not only an attempt at description and understanding with respect to the kind of observation, reflection and participation that we are referring to, but the opportunity to experience these modalities as a formative process of opening ourselves to a more holistic consciousness. To this end, from the end of this chapter onwards, each chapter will close with an exercise or a number of related exercises (or practices) which might be used to cultivate the new faculties which Goethe talks about.

These practices are all, in a broad sense, artistic. For it is through the practice of art that the imaginative faculty is exercised. 'He to whom Nature begins to reveal her open secrets feels an irresistible longing for her worthiest commentator, Art', said Goethe.[3] Rudolph Haushka, following Goethe, notes that 'artists *live* in the objects of their study and create them freshly and revealingly again. This type of creating calls for something that goes beyond trained hand or eye: it requires vivid activity of soul and spirit.'[4]

However, in the narrow and specific sense of the word artistic – that is, as representation and interpretation through the medium of line, colour, form, sound, language and so on – they are not all artistic. Many of these exercises are geared towards developing the art of living itself, as a means of cultivating the ability to read the meaning of what surrounds us. These exercises, in fact, comprise three different types. First, 'finger' exercises (of the kind that pianists might do before sitting down to play) – literally exercises in observation of the world which are intended to build capacity for such observation; they do not require you to have had any artistic training or even demonstrated aptitude, though, as with all these practices, they do require commitment. Second, self-reflection exercises – done alone or in pairs or small groups, these are intended to help you access your own processes of development in order to enable you better to appreciate those which you attempt to read in others, and also to help know yourself well enough to cope with possible projection in the act of observation. Third, meditative practices of various kinds which are designed to strengthen the thinking and imaging processes themselves.

Only one of these types of practices will follow each chapter, though there may at times be more than one exercise included within that particular practice. The practices will follow the general gist of the chapter itself – not in the sense of learning to do what the chapter has focused on (which would be a reductionistic approach) but rather to enable some form of deeper experience of, and engagement with, the complexities which the chapter has sought to describe and understand. The practices do not have to be undertaken in the order in which they appear; neither do all or indeed any of them have to be undertaken at all. They are indications only; invitations, if you would. They can be done formally – as in setting time aside – or informally, as in attempting whichever of them feels appropriate as you are working with a social situation.

There are many such exercises and practices, and you may know of, or discover, others which work better for you. There is no doubt that the value of this book does not depend in the first instance on whether or not you engage with these exercises. Equally though, there is no doubt that the development of the faculties which we are referring to is dependent on disciplined practice. The task of honing one's abilities as a development practitioner takes years, and requires exercises of one kind or another. One can develop such abilities simply through constant social development practice and particularly reflection on such practice. But exercises such as those included here go a long way to deepening and accelerating the transformation which is required of every social development practitioner.

EXERCISE

Drawing

At the outset, it helps to draw. Whenever, on outings, in meetings, through drawing classes. With some good paper and some good drawing pencils, we can dedicate ourselves to the task for a while. There is nothing which better develops the ability to pay attention. Drawing is nothing more than the ability to see; it both develops the capacity as well as derives from it. Forget about finished product, about embarrassment, about wishing you were a 'real artist'. For those who are, this section is irrelevant anyway, and they start the development of new faculties at a decided advantage. But they are merely paying attention. These drawing exercises have to do with the process of seeing and drawing itself; the product is for your eyes only. I have found drawing exercises invaluable because I always believed I could not draw (and still have my doubts). But I have learned to see through drawing (partly), and I have derived endless pleasure. (Goethe said that he could only claim to understand something when he could draw it from memory. This is way beyond our ambitions here, but the value of the practice is clear.)

Here follow four exercises in drawing. The guidelines are very truncated – there is little space and I am not a drawing teacher. There are invaluable books which you may refer to for these and many more drawing exercises.[5]

Find a relatively complex line drawing done by an artist, and turn it upside down before looking at it.[6] Perhaps get someone to do this for you; it is important that you do not see it right way up. Because this exercise will help you to see afresh, by showing you that if you do not conceptualise and configure what you are looking at before you draw, if you draw the lines as you see them – as lines – and not as representing a hand or a nose or a flower or whatever, then what you draw will turn out to be amazingly accurate. You will discover that you can draw, though you had always thought you couldn't. And all because you are paying attention to what you are seeing – lines on a page – rather than to the meaning which you would organise in your mind. Simply draw the lines that you see, sitting down unhurriedly to the task. Never rush – you must take the time to become immersed in the experience.

Here is another exercise called blind contour drawing.[7] Draw the contours, the outlines of objects, complex objects – your own hand, a flower, a hat, the bark of a tree or the contour of a stone – without taking your eye from the object. Do not look at what you are drawing on the page, but plunge into your seeing of that which you are drawing, and it will come alive, and make apparent its relationships to other aspects of itself and to its surroundings. The object will begin to glow and speak its own particular language and meaning. Draw slowly, and slowly you will enter into intimate relationship with what you are drawing. Most important is the frame of mind into which we enter when we engage in this exercise. If we do not hurry, but take our time and do it slowly, we enter into a meditative, calm, reflective state in which our consciousness spreads out and seeps into the world; a state of centred clarity of heightened awareness, in which perception is both focused and diffuse, as if we were an eagle on the wing, able to encompass its entire surroundings and yet focus in on the tiny spot of moving mountain hare at the same time. It is this state of awareness which is the point of such exercises. We can come to appreciate and recognise it so that we may enter into it at will; it is the basis of a holistic way of seeing.

After you have done a number of these drawings, over a number of days, begin modified contour drawing by glancing every now and then at your drawing, as you draw; perhaps begin to use an eraser, and have some concern for final product (but only in terms of your relationship with the object drawn).

Try particularly to observe and draw plants – parts of plants as they grow and change, follow the gestation of a blossom, the unfolding of a flower, the ripening of the fruit – draw them in their process of metamorphosis. As a further step, try to depict the process of metamorphosis itself. For a start, take a sheet of paper and shade it in charcoal, from light above to dark below. Take a charcoal eraser and, beginning with a seed in the middle, allow yourself to unfold a tree, moving alternately above and below the earth as the tree spreads itself out. Experience the process of development and change as it evolves out of you.[8]

Then, try to depict the process of metamorphosis and change without using a tree, or any living representation. Simply use lines, form, colour. Allow the inner feeling to emerge on to the paper however it will. Through these exercises we begin to gain an appreciation for the process of development and change itself, and the whole movement underlying living processes of change will begin to make itself felt within, so that it may be comprehended without. Similar exercises can be attempted with other such 'invisible dynamics', such as power, joy, depression, particular relationships; even one's intuition of another person's character, or the character of a specific situation.[9]

6
Revisiting the Whole

For every atom belonging to me as good belongs to you.

Walt Whitman

There are two further considerations which must be raised concerning the whole.

We have seen that the whole is of a different order from the parts, and that we cannot see it from within the same mode of consciousness that we do the parts. We see things through the organ of sensory sight, and we see wholes through the organ of imagination. And we can use both simultaneously.

It is to this notion of simultaneity that we must refer briefly now. It is difficult to conceive, given the logical frame of mind in which we are steeped, that opposing things can happen at the same time, and in the same space, without contradicting each other. Yet the realm of logic concerns mostly the world of inanimate things, of solid bodies forever external to each other. It has been extrapolated to hold universally, but life itself does not follow the rules of logic, and in order to move towards holistic seeing, or intuition, we have to release ourselves from the fetters of logic, and embrace seeming contradiction as not only possible, but normal.

This applies directly to our understanding of whole and parts. We have talked of the whole variously as either *underlying* the parts or *emerging out* of the parts. (And for the whole we could substitute the concepts 'living process', or 'meaning'.) Yet we talk in this way because language itself is a function of the analytic mode of consciousness, not necessarily because it is accurate. In fact, it is slightly misleading. The whole does not underlie the parts, neither does it emerge out of them; *it exists simultaneously with the parts*.

Our normal way of thinking (as with our language itself) is a linear one: one thing follows, or is caused by another. It is easier, then, for us to think that the whole either precedes the parts – in which case the parts are determined by the whole, defined by it, and so are somehow subservient to it – or that the parts precede the whole, in which case the whole is built up out of the parts and is dependent on them. While both of these perspectives have some validity, they do not embrace the entire truth, and they subscribe to both a linearity of thought as well as to the risk that we may see the whole as just another part, albeit a *super-part*, or a connecting part. But the whole is of a different order from the parts, and neither precedes nor follows them but exists simultaneously with them, and manifests as they relate. We may speak of the whole as being *enfolded* in the parts, and our task as to help unfold what is enfolded within.

Take the example of reading, which was mentioned in Chapter 4. There we saw that we could see both the individual marks on the page and the meaning of the word at the same time – what we called twofold seeing. Now, it would never occur to us to suppose that the meaning was *behind* the letters of a word, or underlying them as if the letters needed to be explained by reference to something behind the word. It is obvious that both letters and meaning belong to the word at the same time, but that we experience (see) them differently. Such is the concept of simultaneity, which is so very difficult for our everyday consciousness to accept, as it goes so much against the grain of our usual way of thinking. Yet it is mentioned here because an ability to embrace the concept (and practice) of simultaneity takes us a long way along the path of seeing differently. Learning twofold seeing is learning to see both foreground and background simultaneously; meaning is enfolded in the relationships emerging between the specific parts.

The second consideration concerns the nature of the whole as an expression of change and becoming. At least as far as organic wholes are concerned, and the holistic perception of living process, we should not conceive of the whole as something static and fixed. The organic whole is never still, is never fully formed, but is always in a state of becoming. It is always in movement, and when we apprehend such a whole it is always movement which we apprehend, not a state of rest. The whole, indeed, *is* process.

The I Ching, that ancient book of Chinese wisdom, says that 'Nothing is absolutely at rest; rest is merely an intermediate state of movement, or latent movement.'[1] This resonates with the new sciences when they note that what we think of as things, as discrete particles or particulars, are actually intermediate states in a constantly changing network of interactions and relationships. Discrete particles come into existence often only temporarily, when invisible fields intersect. Goethe notes that it is a mistake to assume a congruous whole which is determined, completed and fixed in its character. He writes: 'If we consider *Gestalts* (wholes) generally, especially organic ones, we do not find anything permanent, at rest, or complete, but rather everything fluctuating in continuous motion.'[2] If we so wish to use the concept of a whole for something static, he goes on, we should think thereby 'of something held fast in experience for but an instant'.[3]

The whole, therefore, is always in a state of change, of metamorphosis. Indeed, the organic whole may be described as the process of change, or the state of metamorphosis and becoming, as it is expressed by a living organism at a particular moment or in a particular phase. Marcus Aurelius notes that 'all things are continually being born of change ... Whatever is, is in some sense the seed of what is to emerge from it.'[4] And Goethe again: 'What has been formed is immediately transformed again, and if we wish to arrive at a living perception of nature, we must remain as mobile and flexible as the example she sets for us.'[5]

This is why exercises in observing and drawing the metamorphosis of plants and parts of plants are so valuable, and in fact formed the cornerstone of Goethe's work. Take the image shown in Figure 6.1.[6]

If we study the metamorphosis of a leaf in this way, we can begin to penetrate to the movement, the process of becoming, as an invisible whole which manifests as physical matter – as leaf in this instance – from time to time. The individual leaves, in this case, are the particulars – the parts – and the organic whole is the movement which unites and connects them, the invisible field which is enfolded within them, and of which they are static manifestations at different points in time. The landscape mentioned in chapter 5 can also be observed and understood in this way, as can the relationship between water, rock and surrounding vegetation in a river, or the unfolding of a human life, or a social situation. In this way we can begin to see process, the living whole, directly.

This has obvious implications for us as social practitioners attempting to read and understand the invisible processes, connections and fields which are at work in social organisms at the time when we are seeking to intervene – a point we expand upon in the following part. For now, as with the previous point, it is mentioned in order to help in our transition from an analytic to a holistic mode of consciousness. Learning to read the whole as movement rather than as static object assists in moving from a 'thing' mentality to an appreciation of process. In doing so we move towards a certain sensibility.

Figure 6.1   Diagram of a leaf

What may it mean, this ability or inability to see wholes, for the way we are in the world?

Consider the salmon, which migrate across the oceans of the world, or the turtles which do the same, or certain species of butterfly, or the flocks of tiny birds, like swallows, who move from north to south and back again, always in search of summer, always returning to the same roosts which they inhabited the season before. The mystery and miracle of life. We have referred already to the intangible morphogenic fields within which they are able to complete their journeys unconsciously, as it were, guided by the invisible thread which patterns the species' behaviour. It is impossible to imagine them planning the journey step by step, one stage at a time. They must all see the journey as a whole, as one, and they themselves as an integral part of it, the journey and the traveller comprising one living process. For if one tries to see it in steps, the miracle becomes inconceivable.

A form of mastery indeed. But sentimentality will hide the other side of the story. For these travellers are trapped, like the spider spinning its web, the kingfisher harvesting its fish. All are so specialised as to be trapped in morphogenic fields from which they cannot escape. Not that they would wish to – the whole is so complete a circle that player and play are one, and breaking the boundary of the play would be incomprehensible and absurd.

But we, both blessed and cursed with a form of consciousness which allows us to reflect on ourselves – self-consciousness – have broken through, we can manipulate our world, our circumstances, ourselves. We can change the play, and are free – to a degree at least, for underlying patterns still inform much of what we do. But how much we must pay for that freedom? Perhaps the ultimate price is the loss of the capacity to experience the whole, to be participants and players rather than observers and onlookers. There are two sides to our situation as well; our very freedom can trap us outside in the cold, beyond the warmth of life's hearth. We have to relearn, consciously this time, to become participant and to see the whole.

Perhaps it was not always this way. When tradition and instinct and the sense of fate were stronger, then our sense of belonging provided us with warmth and purpose, and we participated in the wheeling and turning of the cosmos like all life's creatures. Consider this statement from Smohalla, of the Nez Perce Native American people, on being challenged in his way of life: 'You ask me to plough the ground. Shall I take a knife and tear my mother's breast? Then when I die, she will not take me to her bosom to rest. You ask me to cut grass and make hay and sell it and be rich ... But how shall I dare cut off my mother's hair?'[7] Such is one expression of an original way of being. But modernity (and post-modernity) is the expression of an altered way of being. It has shattered the old gods and even the new, and left us free, but in an echoing void, without centre or given direction. Rudolf Steiner, spiritual philosopher and teacher, wrote: 'The stars spoke once to man, but they are silent now.'[8] And the literary critic George Steiner speaks of 'The promise of an eerie freedom.'[9]

In this new-found freedom we have largely lost our spiritual connection, and are sunk deeply into matter. Like drunkards, we bumble blindly amongst outreached objects, oblivious at the core. As put so nobly by W.B. Yeats:

> Now that my ladder's gone
> I must lie down where all ladders start,
> In the foul rag-and-bone shop of the heart.[10]

It is time to take what we have learned from the experience gained in manipulating the world and move on. Take that ability to step outside, and to act on the world as if it were a thing – to construct and control – and offer it back to that part of ourselves which knows, utterly and without words, that we are participants. So that we may become *conscious* participants.

As individuals, we face a similar task as we grow through the stages of our lives. Consideration of these stages will provide us with further illumination. Psychologists note that there are two major modes of organisation for a human being: the *action* mode and the *receptive* mode.[11] As infants, we are in the receptive mode, but this is gradually dominated by the development of the action mode, formed through our interaction with the physical environment. Through our manipulation of solid bodies we gradually develop the ability to focus the attention and perceive boundaries – to discriminate, analyse and divide the world up into objects. A necessary shift, but as a result we become biased to perceive selectively only some of the possible features of experience.

The alternative mode of organisation, the receptive mode, is one which allows events to happen, rather than seeking to impose and manipulate. Instead of being verbal, analytical, sequential and logical, this mode of consciousness is non-verbal, holistic, non-linear and intuitive. It receives, rather than attempting to control. The development of wisdom and maturity in the individual human being is partly a function of the gradual coming together of these two modes; either one on its own leaves us incomplete and, to a certain degree, trapped. The receptive mode – the realm of imagination, wholeness and belonging – must be re-learned once the action mode has become dominant, and the two integrated into a form of simultaneity or twofold seeing.

Perhaps, in our development as a global community, we have arrived at a similar juncture to that experienced by us as individuals, and need now to re-learn and incorporate a way of being which has become lost as the horizon from which we set out has disappeared behind us. Certainly we have become almost exclusively analytic in our approach to the world. Our tendency is to deny the whole, to take things piecemeal, and thus we avoid seeing the manifold connections in what we do. We dissect, and analyse; we seek to control and manipulate with little regard for sustainability or consequence, for the integrity of the whole. And not merely with respect to science and ecology; but politically, socially and economically. The history of the social for the last few hundred years, and particularly in the century just passed, is a history of manipulation, oppression and extraction, and especially division – the separation of us and them. The (literally) unbearable tendency to control and manipulate, without regard for the wholeness of the social fabric. Underpinning the history of colonialism, communism, and the current excesses of capitalism is the same instrumentalist tendency, the same control mentality, the same reductive myopia. And the social fabric, like so much else, has been torn asunder.

We live now within the tidal pull of elite globalisation, unfettered capitalism, and instant communication. These have their advocates and critics, but the tendency is to focus in on specific instances of excess or advantage, injustice or prosperity. It is only recently that (still too little) attention is being paid to the fact that we are all being swept along by an invisible current about which we have little awareness, and even less control. The institutions which power the gravitational pull of economic hegemony are substantial beings – organic, living wholes – in their own right, beyond the reach of most individuals, and even those who head them are held in the sway of their path. We have released forces which are manifesting as fields, even as entities, immense beyond our ken. We are all caught within living wholes, within fields of force, which we have released, but which we deny the existence of because they cannot be directly seen. The concept of countervailing power, and the rise of cultural power (through the action of civil society) as a countervailing force – as articulated, for example, by Nicanor Perlas[12] – marks the beginning of a recognition as to how these fields of force work, and how they might be guided. Great benefit would be derived, however, by a more conscious acknowledgement and awareness of how such beings and wholes, such spiritual forces, emerge as influential fields which pattern our everyday activities and responses.

The story of electricity is instructive. According to Ernst Lehrs:

> [W]e need [to] compare the present relationship between production and consumption in the economic sphere with what it was before the power machine, and especially the electrically driven machine, had been invented. Consider some major public undertaking in former times – say the construction of a great medieval cathedral. Almost all the work was done by human beings, with some help, of course, from domestic animals. Under these circumstances the entire source of productive power lay in the will-energies of living beings, whose bodies had to be supplied with food, clothing and housing; and to provide these, other productive powers of a similar kind were required near the same place. Accordingly, since each of the power units employed in the work was simultaneously both producer and consumer, a certain natural limit was placed on the accumulation of productive forces in any one locality.[13]

The discovery of electricity changed this relationship profoundly, for 'electricity is distinguished from all other power-supplying forces precisely in this, that it can be concentrated spatially with the aid of a physical carrier whose material bulk is insignificant compared with the energy supplied'.[14] In other words, limitless accumulations of power can now occur in one place.

Natural limits, therefore, have been all but swept aside. We still talk glibly about sustainability, but without really being conscious of our own complicity with the forces which mould our lives. Such forces gain in stature and authority until they become commonplace, go unremarked, and we could not conceive of living otherwise.

There is no question here of denying the value of electricity. What is being suggested is that we become compliant with what has become commonplace, and thereby become trapped within forces, patterns, fields of which we are unconscious. We become so trapped because we *are* unconscious. Unobservant of the wholes within which we live, and which we create. Initially ephemeral and fragmented strands of process gradually cohere into the landscape within which we walk unwitting. Or, as James Hollis puts it: 'Through our ingenuity we have made things of great power, and now we serve those things.'[15]

Hollis goes on to say: 'to be modern is not just to be alive in this era but to understand what most characterises our Zeitgeist, namely the erosion of that invisible plane which supports life on the visible plane'.[16] Our urgent task, then, is to learn to see the fields which, like morphogenic fields, dominate our lives, and leave us, paradoxically, unfree to use our new-found freedom. Only in this way will we be able to choose a preferred future, and act judiciously to accomplish it. We cannot deny electricity, or the forces of globalisation, but we can become aware of the invisible fields which they create, and so begin to operate with a degree of selfhood, of independence, of conscious choice, with respect to even the smallest of our actions. Reading of the landscape is a prerequisite for finding a path through.

We do unto ourselves what we do unto others. This is not the logic of solid bodies, but the indubitable consequence of wholeness. It is the logic of the organic, of participation rather than observation. The capacity, or the lack of capacity, to see holistically reaches into every sphere of life. Lest we create morphogenic fields which cannot sustain us into the future. As Christopher Fry wrote:

> Affairs are now soul size;
> the enterprise is exploration into God.
> It takes so many thousand years to wake,
> but will you wake, for pity's sake.[17]

Nothing is fixed, everything is in movement, in process, all the time. We are participating in an ongoing adventure of the spirit. No longer are we onlookers observing a cold, mechanical world of inert matter and isolated events, supposedly reducible to calculable laws with which we can (theoretically) predict and control. We become active adventurers in the journey of becoming. The world is mobile and fluid, constantly in flux, and unpredictably responsive to our (even inadvertent) touch.

The world clearly needs our self-consciousness as human beings to be able to manifest towards its full potential. The journey of human consciousness, and the journey of the world's evolution, are inseparable. We have to cultivate the capacity to understand our own experience. The more we understand, the more there is to understand. Our understanding enables our world to emerge more fully, to realise itself. The more we see, the more there is to be seen, for the bigger we become, the more there is to the world. To be alive is to participate in the unfolding of a great mystery. The world's unfolding and our own creativity are one. We are implicate; and we are also held. Nothing is fixed. We pattern the future through the way we are, and we take the way we are from the patterns which emerge.

EXERCISE

Listening (1)

In the quest of accurate observation, the art of listening is as important as seeing. We will explore listening to individual others, and to social situations, in a later exercise on listening. For now, it is the faculty itself which we attend to. As with seeing, we listen and listen and yet – to paraphrase Goethe – we risk listening past a thing. We look here for an attentive rediscovering of ordinary physical hearing as a means of developing a faculty which might assist us to move beyond.

If we listen to various things with sufficient attention then we begin to penetrate the depths of things, into their infinite richness, in ways that may not have been apparent before. Listen first to individual sounds emanating from different sources – from a knocking on wood to the rushing of wind to the clackety-clack of a train to the song of a bird. Listen for the different ways that different kinds of wood sound when knocked upon. How does this compare with metal, or stone? Compare the sound of rain on water to rain on a roof; compare the sound of wind to rain, and the different sounds that each make as they travel through different vegetation. Listen for the difference between animate and inanimate, between the elements and animals, between animals and birds and insects. All the sounds coming from the inanimate differ from each other, but also from that which comes from animals and humans, where the soul quality begins to emerge. Listen for that soul quality – it will begin to express all the nuances and character of different situations and dispensations. A pig being slaughtered compared with a pig squealing for its share of food. A baby crying and a baby laughing.

If we listen closely, as disciplined exercise rather than as superficial and unconscious habit – consciously – then whole worlds open up. Large fields of hearing which had previously been denied begin to emerge. We discover that we can begin to listen past our immediate response, for the character which echoes out of the depth of that which we are listening to. And when we go back, from the animal and human world to the inanimate world of objects again, we hear these objects with new depth, and other worlds are revealed. (We may discover a new accuracy if we also, when opportune, listen without seeing.)

A further listening exercise – nice to do while walking – is to listen initially for one sound only; focus in on the dominant sound. Gradually, then, begin to allow other sounds to enter, keeping them separate from the original and from each other, so that eventually one hears many sounds active at the same time, yet all discrete and all being, as it were, individually paid attention to. Try to expand your listening in this way to include more and more sounds, those which are fainter or further away and more difficult to hear. Try to expand your listening out into the periphery of itself. The San of Southern Africa imagined 'turning one's ears back to the souls of one's feet, to listen for what is in the air'. The infinity lying at the heart of the world gradually begins to manifest, as your powers of listening improve.

All these exercises demand a certain commitment, a certain discipline, and a certain trust that something will begin to emerge out of nothing. Even if such certainty is still tentative, if you accept the paradox, and practice is infrequent and sporadic at the beginning: if the exercises are to be of any value, active engagement is fundamental.

Part II

Understanding

In the last analysis the world is a system of homogenous relationships – it is a cosmos, not a chaos ... The ultimate frame of reference for all that changes is the nonchanging.

I Ching (Book II)

Don't be confused by surfaces; in the depths everything becomes law.

Rainer Maria Rilke

7
Freedom and Constraint

Consciousness ... is of a much higher order than twice two.

Fyodor Dostoyevsky

As social practitioners, we learn to see so that we may understand the social organism with which we must interact. And understanding means developing the ability to apprehend the whole, the invisible fields which form the manifest organisation. To paraphrase E.E. Pfeiffer, with reference to the earlier quotation (pp. 20–1) – wherever organisation exists, it takes form. The formative forces which give shape to organisation exist as intangible fields which pattern and aggregate the material (tangible and manifest) aspects of organisation. Without such forces no organisation can exist, and without recognition of the existence of the formative principle in organisation one cannot understand organisation itself.

As a social organism grows and develops, it manifests its own uniqueness, but this pattern is based on what may be termed archetypal templates – the patterns of organisation which live deep within all social organisms, as bedrock.

As we have already noted, everything is in a state of movement, of change. Nothing that is alive is not changing. Yet this does not mean that all is chaos, that change is arbitrary, unique and incomprehensible. The I Ching, that ancient Chinese text which has as its entire rationale the understanding of change, notes that the ultimate frame of reference for all that changes is the *non-changing*. It goes on to say: 'If we know the laws of change, we can precalculate in regard to it, and freedom of action thereupon becomes possible. Changes are the imperceptible tendencies to divergence that, when they have reached a certain point, become visible and bring about transformations.'[1] It takes a fine discrimination to apprehend the 'imperceptible tendencies to divergence' upon which our art is premised. Such seeing needs to be backed up by an understanding of the 'laws of change'; some frame, or story, is necessary against which we may make sense of what we are seeing. Strangely, our freedom is not compromised by such laws, but enhanced when they are recognised, respected and engaged with.

Joseph Campbell, the celebrated explorer of myth and its relationship to conscious living, once remarked:

> People say that what we're all seeking is a meaning for life. I don't think that's what we're really seeking. I think that what we're seeking is an experience of being alive, so that our life experiences on the purely physical plane will have resonances within our own innermost being and reality, so that we actually feel the rapture of being alive ... Myth helps you to put your mind in touch with this experience of being alive. It tells you what the experience is.[2]

It is in this sense that we look in this section to explore what may be termed the underlying archetypes of change.

These paradoxes – between freedom and constraint, between change and nonchange, between universal and unique – which characterise the movement of the social organism as it develops and changes, can perhaps best be understood by exploring our current condition as individuals. After all, social organisms are created by us, stamped with our own sensibilities and confusions. The organisations, communities or social situations we create are created in our own image. We have to know ourselves in order really to penetrate those things that are created by us, those beings (new wholes) who are enabled by us to emerge. The patterning of social organisms closely mirrors our own patterning. And when we look closely at ourselves we find ourselves permeated by paradox.

No longer do we look on stars and animals with the sense of belonging which characterised the San of Southern Africa; no longer can we sing the songs of creation which keep the earth alive and growing, as did the Aboriginals of Australia. We have moved a long way from that sense of participation and oneness. We have learned to see the world from the outside, as it were; objectively, so-called. We have learned to act on the world (and on others) as if it were entirely separate from ourselves; we have learned to manipulate, to esteem control and predictability. We have learned to define boundaries, and always to put ourselves on the other side of the line. Our project has become almost entirely focused on emancipation – from tradition, from given cultural norms, from social expectations, from ecological and environmental limitations; most relevantly, from the archetypal fields which pattern both the natural world and human existence. In the process, we have cut ourselves off from these wellsprings of life, have alienated ourselves from the very pulse of the world. But we have found a form of freedom, freedom to choose, to act without constraint. Or have we?

That great poet of the human condition, Rainer Maria Rilke, in his seminal work *The Duino Elegies*, captures the dilemma in two different moods.

> ... how he was tangled
> in the spreading
> roots and tendrils
> of inner event
> twisting in primitive patterns
> in choking growths
> in the shapes
> of killer animals ...[3]

Taken from the third elegy, Rilke brings to expression some impression of the invisible fields, or archetypal patterns, which infiltrate and influence the movement of a human life. As he goes on to write, such archetypes are 'already dissolved in the water that makes the embryo float'.[4] However, in the very next elegy we read the following:

> O trees of life
> when is your winter?
> We're not in tune
> We're not instinctive
> like migrating birds.
> Overtaken
> overdue
> we push ourselves suddenly
> into the wind
> and arrive surprised
> at an indifferent pond[5]

Which reveals a different facet. The pond, which once resonated with sympathetic meaning, we and it belonging to each other, is now indifferent. And we are surprised, not quite sure where we are. It is not, however, that we are suddenly free of all morphogenic fields, but rather that we are both free and not free, in an uncomfortable place of liminality, of transition. Rilke's following words read, 'We understand blooming and withering/we know them both at once.'[6] Precisely. We are simultaneously inside and outside, participant and observer, free and constrained, emancipated and alienated. This is the doom of the modern: to be in-between, seeking a new home as the horizons of the old disappear behind us. And as we move out beyond our tatooed past and into an unknown future, we no longer even know whether we are searching for a new home or whether the very concept of home has been left behind.

We used to be born into a particular tradition and culture, and the way would be given us – whether that way was the participatory flexibility of the hunter-gatherer, the feudal rigidity of medieval hegemonies, the ethical code of fundamental religiosity, the presumptuous morality of colonialism. There have been many, many different ways, but, for any individual, they have always been given. No more. As Yeats put it: 'Things fall apart; the centre cannot hold; mere anarchy is loosed upon the world.'[7]

There is nowhere any longer to turn – we have only ourselves. The ancient and classical drama of Good and Evil is something of the past; now, we all have the forces of good and evil within our own breasts. God is not dead – necessarily. But neither can we rely on an external God, as we have done in the past. The power of the future lies within ourselves, even as each of us has our portion of the divine within – as Rilke wrote, 'because inside human beings is where God learns.'[8] We hold the world in the palm of our hand.

There is a wonderful story which comes from a village in Botswana, concerning some youths who wanted to challenge the wisdom of the village elder. The leader of the youths caught a small bird, held it tightly concealed in his hand, and went to the elder with the challenge – 'You, who know so much, tell us whether this bird is alive or dead.' The idea was that if the old man answered that the bird was dead, the youth would release the bird, to prove him wrong. And if the old man answered that the bird was alive, the youth would crush the bird in his hand, to prove him wrong. There was no way that the old man could win such a contest. The challenge then being put to him, the old man stared into the youth's eyes and said, deliberately, 'The answer lies in your hands.'[9]

Just so – the answer lies with us. And the question is not one we can avoid asking, for the focal point of modernity is that it persistently puts the question to us – painstakingly, inevitably, continuously. Our current plight is such that everything we do partakes of light and dark, good and bad. There is no purity of action anymore. As with the dilemma of the dam with which this book started, every action taken in the social realm is fraught with an insoluble ambiguity. Every developmental solution creates a new problem. How do we deal with beggars, how do we deal with the unadulteratedly wealthy; how do we deal with minorities, and with majorities; how do we reconcile the conflicting demands of ecology and poverty, of sustainability and development; how do we reconcile progress and culture? There are no answers anymore – only messy, indeterminate situations which we are called upon to address, every moment of our lives.

Are we fit to meet the challenge? In our haste for emancipation, we believe that we are free of the constraints of antiquity, free to make independent choices. And to an extent we are – this is the privilege gained through the sacrifice of a sense of given belonging and through facing the risk of alienation: the price of freedom. Yet, as Rilke has so succinctly put it, we are not entirely free – those invisible fields run through us still. We have been given the possibility of freedom. We are able, through the gift of (self) consciousness, to step outside these fields, to step outside the whole, and choose. But unless we bring the deep underlying archetypes to consciousness, we remain in their sway, and then our freedom is but a chimera, a conceit, and a danger – for we have lost constraint, but our actions are still dominated by unconscious forces. We are in a state of transition, and we have to pay attention to both our past and our present if we are to guide ourselves towards a responsible and free future.

The concept of 'archetype' derives from the work of Carl Gustav Jung, and refers to the fact that all humans possess a similar psychic structuring process; in other words, that the process by which people develop has certain patterns which recur through the ages, from individual to individual and from group to group. These patterns can be seen as morphogenic fields which deeply influence the way we think, feel and behave. In this sense, they are invisible formative forces which impose pattern on chaos, provide life with depth and meaning, and enable our manifested lives to emerge and develop. Such archetypes lie deep, and are not invented by our consciousness – they lie richly layered within the collective unconscious which is shared by us all.[10]

Fritjof Capra, exploring the meaning of developments in the new sciences, claims that all living systems comprise three essential aspects, or criteria: pattern, structure and process. *Pattern* (form, order, quality) is 'the configuration of relationships among the system's components that determines the system's essential charac-teristics'.[11] Put another way, it is the *design* of the system; that which gives it its singularity. In Pfeiffer's words, mentioned earlier, this may read as 'the formative force which gives shape ...'. The *structure* of the system is the physical embodiment of the pattern, or design. Thus the structure enables the invisible to manifest as visible in the physical realm. *Process* is the continual embodiment of the system's pattern of organisation. In other words, process is the developmen-tal means whereby invisible pattern becomes material manifestation. It is process which differentiates living from nonliving systems; process is the mediation between form and matter.

With respect to social organisms, then, archetypes can be seen as 'design principles', deep formative fields which pattern the organism as it develops its physical identity and structure. Process is the manner through which such embodiment takes place. Our task, as social practitioners, is to work with that process, to enable it to move freely and find its way when it becomes lost or blocked, as can happen with self-conscious organisms blessed and fraught with the dangers of free choice. We learn to read the whole, the invisible essence, because it is here that process is found. And process is not easily read unless we understand the design principles, the archetypal fields which imprint their patterns on the formation of the organism as it grows and develops.

This section is concerned with these patterns. Three major patterns are outlined through the following three chapters. These encompass the dominant fields within which social organisms develop. Thereafter, in the next three chapters, we explore three groups of what may be termed minor patterns. These are specific tendencies which also act as formative fields, but which take place against the backdrop of the three major patterns. Together, under-standing of these patterns coupled with a new way of seeing will enable the invisible and particular dynamics of developing social situations to reveal themselves.

EXERCISE

Developing the Core

The quest for freedom takes us beyond ourselves, beyond the limited horizons and constraints of our everyday selves. Towards a centre of calm and meaning and purpose; towards our own authority, the leadership within, lying at the core of our being. Accessing this higher Self enables stability and flexibility to reign within (therefore without). It is this core which we access when we begin to observe and intervene intuitively and accurately.

We can help ourselves to access this core. First, solitude is needed, short periods set aside daily or at longer intervals, when we can withdraw into seclusion. These are meditative spaces, times for strengthening thinking and imagination, for reflection and contemplation. Many of the exercises which follow require that such space be set aside. Time spent in this way immeasurably enhances and improves our doing in the world; such doing then becomes more authoritative and responsible, more creative.

In this particular exercise, the time spent in seclusion is used in the following way.[12] To think back to a situation, a difficult situation, in which one was somehow tested, challenged, confronted by the circumstances of one's own life. And to look at that situation from a higher vantage point, as though we were a different person. As if we had spent a day working on a plot of land amongst many others, and in the evening had climbed a nearby hill which overlooked this land, so that all the gardens and gardeners, our own included, were laid out so that we could survey the whole area at once. Proportions and relationships change as we gain such vantage. The very act of placing ourself in this position in relation to our life awakens a faculty which no one can awaken for us.

In looking from this vantage point at the situation you have chosen, try to discriminate between what was essential in that situation, and what was inessential. This is difficult, almost meaningless; everything had to happen as it did, so everything seems essential. One way of penetrating the essential is to ask: What forces for development have been formed as a result of the process? And, which are the patterns that repeat, the relationships that I recognise; and where may they be turning? In doing this exercise we are in search of the wellsprings of our very destiny. We begin to discover new depth in our life, new reasons, new ways of reading our own story; a renewed sense of responsibility.

After a while, we can begin to extend the situations we look at, so that we are taking chapters of our lives, significant years or passages of time, sojourns in particular places, the trajectory of particular relationships, and so on.

The more we work in this way, the more we understand of ourselves and our life patterns and choices. A sense of clarity and overview enter our lives. We are also able in this way to understand better the lives of the social organisms with which we work, and to observe their stories with greater accuracy and intelligence. We begin, not least in our own lives but also in the lives of others, to be able to intuit the essential and distinguish it from the non-essential.

8
Balancing Heaven and Earth

A great mutual embrace is always happening
between the eternal and what dies, between essence and accident.

Rumi

We know that the whole is never static, that it is constantly in a state of movement. It *is* that movement; the whole is the unfolding and flowing of a living system in its process of development. How does such movement occur, and why? What is the underlying dynamic that powers the changes; what are the patterns that give it form? For such movement is not arbitrary chaos – though it may look chaotic when seen in the short term, and from too close up. If we stand back and look through the discrete and visible instances of change we might begin to see the pattern forming the dynamic. Put another way, what does *not* change; what is the nature of the *non-changing*?

Leading into the discussion firstly from the ancient tradition of the I Ching, we note that the power of Tao, the underlying principle of how things happen in the world, 'is to maintain the world by constant renewal of a state of tension between the polar forces'.[1] Polarity is our first archetype.

The first two polar forces which the I Ching describes, in fact the two upon which the entire Book of Changes is based, are Heaven and Earth. Heaven is described as above, Earth as below. Heaven embodies the creative principle, Earth the receptive. Heaven disposes while Earth nurtures. In some sense, Heaven is spirit, while Earth is matter; Heaven is impulse, Earth is incarnation.

In the interplay between these forces the world changes, grows and develops. The I Ching notes: 'Opposites do not combat but complement each other. The difference in level creates a potential, as it were, by virtue of which movement and the living expression of energy become possible.'[2] The world turns, it seems, on the tension generated between the opposites of spirit and matter, manifesting as creativity and receptivity, impulse and manifestation, sacred and profane, and so on. We are naturally inclined to promote one polarity over its opposite, but it is this very bias which causes distress and stagnation in social systems; a certain fundamentalism and potentially explosive rigidity creeps into the social organism. We label one 'good' and the other 'bad', not understanding that the healthy movement of the organism is dependent on both polarities functioning freely *at the same time*, because only so is the power generated to keep the whole in movement.

There is a profound clash between the Newtonian and Goethean world views with respect to colour. This is interesting because it may deepen our appreciation for polarity. Newton believed – and thought that his experiments bore him out – that all colour was to be found embedded in white light. The different colours, he said, consist of rays differently refrangible – in other words, when white light is passed through a prism, the different coloured rays are separated out because they are refracted through the prism at different angles. This was based on experiments in which white light was passed through a prism onto a wall, revealing the colours of the rainbow. Thus Newton assigned to each colour a mathematical value – its angle of refraction through a prism – and sought a mechanical model as explanation for the different rays, which subsequently evolved into the wave model of light. This theory of colour is still generally accepted today.

However, when Goethe tried to repeat the experiment, he saw something other. He noticed that when white light is passed through a prism onto a white wall, no rainbow is formed – no colours emerge at all. Only where there is dark – a black band, say, on the wall, or a black card against the white wall may be used in such experiments – only where there is dark *and* light do colours emerge. In fact the colours emerge on the boundary between the dark and the light. Goethe identified two colour spectrums: the light spectrum (the reds and yellows) and the dark spectrum (the blues and violets). These form two distinct spectrums, with green emerging at the interface between them.[3] This provides an entirely different possibility – that colour is not contained within white light. If colour emerges anew out of the interplay between the dark and the light, then the purely mechanical explanation of rays of different wavelengths does not satisfy our quest for understanding.

Newton had denied the existence of the dark; for him, dark was merely absence of light. From Goethe's observations it is clear that dark exists in and of itself, as light does. Every colour, every character, every shade of meaning and expression in the world emerges from the dynamic between light and dark, inspiration and shadow, heaven and earth, impulse and manifestation, sacred and profane. We cannot have one without the other, and valuing one more than the other does not serve. They belong together, and determine each other. It is the very polarity between them that forms a new unity, a whole, the next developmental step. Between heaven and earth, between spirit and matter, lies the middle realm of soul, partaking of both, mediating between both, uniquely formed of the particular interplay between both.

However much we may wish to serve one polarity rather than its opposite, that very wish will defeat its end. It seems that life does not permit the domination of one polarity over its opposite; attempts to impose this way on social situations result only in stuckness, distortion and pain. The dynamic through which this pattern (the polarity archetype) manifests itself appears inescapable. The more something is emphasised, the more it turns into its opposite. In *The Tao of Leadership* John Heider states: 'If I do anything more and more, over and over, its polarity will appear. For example, striving to be beautiful makes a person ugly, and trying too hard to be kind is a form of selfishness.'[4] Life, like the flow of water, always seeks a balance; it does not permit of excess without seeking compensation. Through such law all change and movement is processed.

There are organisations in which impulse and creativity are rewarded and respected, where structure and regulation are denigrated; and vice versa. There are groups which respect ideas and ignore implementation, while others focus entirely on action and lose sight of the point of what they are doing. There are institutions too sacred or idealistic to involve themselves in the practicalities of daily struggle, while others concentrate on strategy and manipulation towards a bottom line of self-promotion only. All of these organisms will experience blockages in their processes of development because they ignore the most basic underlying field of all – the energy which arises as tension between opposites. They try, in fact, to rid themselves of such tension by avoiding or rejecting either one of a pair of opposites; the result is either stagnation (death) or crisis (backlash).

There are institutions, originally of the spirit, which so emphasise their perspective on 'the good' that they end up with sets of rules, regulations and injunctions which restrict and diminish the very human freedom which they had been promoting. Those rigid 'religious' hierarchies in which a yearning for the light has turned into a consuming fear of the dark, and there are more laws restricting certain activities than there are practices which encourage others.

Or my own country during the years of apartheid, when a people who prided themselves on their struggle for freedom ended up by creating a highly restrictive society whose rules and regulations cast an oppressive gloom even over the people whom they were supposedly protecting, thus entrenching a reactionary stance in the name of freedom.

There are institutions which have so focused on regulation and bureaucratic procedure that they lose sight of individual workers and thus paradoxically leave those workers free to do what they wish. There appears a contradiction at the very heart of such organisations, as for example the bureaucratic non-governmental development organisation. On the one hand, the development practitioners themselves feel trapped inside a formidable and inflexible bureaucracy – there is little freedom to move, little space to innovate and initiate, everything is performed according to set procedures which are not always logical or thoughtful. On the other hand, there is no common development or field practice amongst them; there is no commonality of profession and discipline – so far as the real work is concerned, the development practitioners are left on their own to make out as they will, *precisely because the excessive attempt at regulating their practice and performance has led to ignorance about the latter through a one-sided focus on regulation and procedure for its own sake.*

There are societies and groups which insist on such radical forms of 'democracy' – call it anarchism – that they deny the need for leadership or coherence, until in the vacuum thus created a shadowy and hidden form of leadership begins to assert itself, and conformity grows without acknowledgement. There are also societies and groups which so insist on focused and accountable leadership and on a coherence of practice and value that participants experience a sense of security which allows for deviance and experimentation.

Too much structure encourages aberrations and lawlessness to manifest; too much freedom turns into a form of licence which allows the powerful to dominate the powerless, and thus curtails freedom. There are many examples from organisational practice, but need we go further than the phenomenon of globalisation which affects our society at large? Both wealth and poverty are being created at unprecedented levels; solutions and problems seem to leapfrog over each other as in an endless child's game; an emphasis on individual freedom holds greater and greater numbers of people powerless. The world's poorest nations are amongst the richest ecologically. Financial power enables the spread of a global culture – rooted in the West – which drives other languages, customs, cultures and traditions into hiding. When all values are subsumed to the economic, as they increasingly are, how much do we lose with respect to social values, to artistic values, to cultural and language diversity, to biodiversity? The world that we are creating with our fixation on the economic (and on economic wealth) is becoming immeasurably poorer with respect to everything which lies outside of this fixation.

Jung regarded fanaticism as overcompensation for doubt, which is a good example of the turning of polarities. The observation penetrates to the heart of the matter. We tend to fall asleep within the lullaby comfort of a particular bias – it is comfortable precisely because it is relieved of the tension between opposites. We often wake when it is too late, when the thing has turned without our realising it, and we wake on the wrong side of the divide. Or the extremes become so attenuated that the situation reaches crisis point and explodes.

The way to work with this particular archetype is always to seek balance. To look for, and to hold, the middle ground. But to seek balance, not compromise. Compromise diminishes each polarity to the point where it is no longer at odds with the other – and then we have no tension, and therefore no energy. Movement and change become sluggish and thick. The situation becomes flaccid. The technique of compromise is to reduce each polarity to its lowest manifestation – and the tension disappears and the world becomes grey.

Finding a living balance is far more difficult. It seems the foundation of creativity and development. It means holding both polarities in the fullness of their power and meaning, and holding them equally, so that the tension is heightened and the energy immense. William Blake wrote: 'Without contraries no progression. Attraction and Repulsion, Reason and Energy, Love and Hate, are necessary to human existence.'[5] So adopting threefold thinking rather than the dualistic thinking which pervades our time.

Every well-formed whole is created through such duality, yet transcends the duality by creating something threefold – the whole is formed not of one or the other of the opposites, but through the relationship which arises between them. The endeavour is exactly to move beyond the dualistic approach into one which recognises threefoldness. Look at the list which follows:

foolhardy – **courageous** – cowardly
inquisitive – **interested** – indifferent
vulnerable – **strong** – impervious
in – **threshold** – out
like – **empathy** – dislike
expedient – **authentic** – principle
light – **colour** – dark
expansion – **centre** – contraction

The richness of the world seems to resonate from out of the third which forms the whole.

We often (and glibly) contrast good and evil, as two opposites or polarities. But what if evil has a twofold nature – and is simply a bias towards one of the dualities – while 'the good' is a dynamic holding of the centre?[6] The words written in **sanserif** above almost induce a feeling of uprightness, that sense of being centred. To be there, the social organism has to be wide awake, sensitive and responsive to every nuance. Lindsay Clarke, in his novel *The Chymical Wedding*, describes this in the following way:

It was about holding together. If we were to find a key to the explosive condition of the world it could only be done by holding contraries together. That *was* the key ... and the holding together could only be done by ... men and women everywhere who were prepared to quake. For quaking was what happened when you endured inside yourself the tension of divisive forces. It was what happened when you refused to shrug them off, neither disowning your own violence nor deploying it; not admitting only the good and throwing evil in the teeth of the opposition, but holding the conflict together inside yourself as yours – the dark and the light of it, the love and the lovelessness, the terror and the hope. And as you did this you changed. The situation changed ...[7]

Holding the balance, in freedom, is the most taxing task of all. Yet also the simplest. It not only encourages centredness but can only really be sustained through finding a centre deep within. To hold polarities in balance is to hold and be held, to be flexible and fluid yet focused, and principled. Demanding and encouraging a deep authenticity. And it renders the organism attuned to the world around it, and to itself. The price of freedom is indeed eternal vigilance. It is within threefoldness that energy can move. The organism becomes mobile, and a certain freedom, coupled with authority, may be attained. The ideal is not the point, so much as the fact that we are all living at various distances from that ideal.

Imagine before you a range of mountains, and focus on their necks and shoulders, those high and narrow passes which allow egress from one valley and access to another, without having to conquer the summit. Such places are cold and windy and seemingly inhospitable; bushes and trees are stunted there, for lack of water and too much wind; we cannot live there, certainly not as we would in the valleys on either side. It is a liminal space, neither here nor there, in between, on the edge. Yet in this space one can see past and future, before and behind; one gains unique insight into both valleys. One is not of either, but in between. One is always slightly uncomfortable, never resting, in a place of transition between worlds. To maintain this edge, this unparalleled clarity, is to begin to work constructively with a formative force which can wreak devastation on the unwary.

EXERCISE

Creative Thinking (1)

Holding opposites together requires strength of character. Paying attention demands commitment and endurance – simply to keep oneself there, in open and honest relationship, requires a fitness of soul. All of the exercises in this book benefit from such discipline. Using imagination as a means of gaining privileged access to essence needs well-formed thinking faculties if we are not to fly off into fantasy. Here follow some means of strengthening thinking, and developing creative thinking.

Some form of meditation is fundamental to the development of consciousness and focus (see the notes and references for wider assistance here).[8] A basic form of Buddhist meditation – performed daily for at least 20 minutes – is to find a sound upon which to focus (say, one's breath, the wind, birdsong, even the noise of traffic). Trying to keep the mind concentrated on such a focal point is impossible for any length of time – particularly at the beginning. The mind wanders through a range of other thoughts before the practitioner even realises that thinking has drifted. Without despair or anger or disappointment, with acceptance and grace, simply turn away from all these thoughts and focus on the chosen sound once more. Without rancour. Over and over and over again – it is the discipline of sitting and doing it which is important, rather than anything you may achieve. If indeed you continue with the practice, a form of silence and space will begin to infuse you, so that you may begin to experience the world as working through you, rather than you on the world. And powers of concentration and endurance, through this very paradoxical process of discipline and letting go, grow immeasurably.

A further exercise (developed by Rudolf Steiner) is to consider an object of little importance – a matchbox, say, or a paper clip or a hair comb – within a space of silent thought, for five minutes at a time. Think whatever you will about the object – what it looks like, what it is made from, the process of manufacture and distribution, its practical and social value, and so on – but think only thoughts about the object for those five minutes; do not allow your mind to stray. Use the same object for some days – you will find that it gets more difficult at the days go on. At the beginning there was novelty which drew you on; as time goes on there is only tedium. Yet it is in the later stages that the exercise's value really becomes apparent. To continue to perform the exercise at this stage requires not only ever greater powers of concentration but also the ability to create new thoughts from out of oneself, for that which was given by the object has been exhausted. As strengthened thinking metamorphoses into creative imagination the exercise moves to another level.

To move beyond abstract and analytic thinking, and adopt a more holistic, intuitive consciousness, is to begin to see directly the 'thoughts' which infuse and build the systems with which we work. We try to look at things not as finished products but as 'coming-into-being'.[9] Practising this, the thoughts which created (or are creating) the thing become articulated as relationships which, through repetition, form the patterns which result in what one sees. If one begins really to look at things in this way, the world gradually comes alive, and one sees things less as a given set of discrete physical objects and more as a fluid, mingling 'coherence' of developing life processes, which have more the consistency of water or smoke than they do of metal or brick. But we can only experience these things for ourselves. Experience can never be granted, it can only be discovered.

Goethe himself called a similar mode of observation 'exact sensorial fantasy'. He would look intently at a phenomenon in all its sense-particulars, and especially at the way the particulars related to each other, and at the way they changed over time. He would concentrate on the 'coming-into-being' of the phenomenon of the whole – much in the manner described above. He would then close his eyes, and retaining a distinct memory of the sense-impressions in their process of change and transition, he would replay the scene in his mind, transforming the sense-particulars into one another, as observed without. Equally important is to reverse the order and play it backwards. By inwardly recreating the transition from one form, or one phase, or one experience, into another, he uses the power of his mind, of his imagination, in addition to merely the ability to perceive sense-particulars. What he was really doing was to endow objective memory, which by nature is static, with the dynamic properties of fantasy, while endowing mobile fantasy, which by nature is subjective, with the objective character of memory. It is the union of these two polar faculties of the soul which gives rise to the new organ of cognition for which Goethe was searching.[10] Combined with active seeing, this has the effect of giving thinking more the quality of perception, and sensory observation more the quality of thinking, culminating – gradually and through disciplined commitment – in a new organ of cognition.

We can perform such thinking activities on natural phenomena (the lifecycles and metamorphoses of plants or animals, the development of landscapes or weather patterns, the relationship between water and forest, between season and plant), or on human and social phenomena (the evolution of a relationship, the development of a conflict, the progress of a group). Both accurate seeing and memorising are important as well as the ability to imagine, to visualise, from out of oneself. It is in the separate honing coupled with the bringing together, the conscious combining, of these faculties that new forms of cognition are born.

9
The Creative Round

Death of earth, birth of water; death of water, birth of air; from air, fire; and so round again.

Heraclitus

So the fundamental formative force which powers the change process of a social organism, and through which its physical manifestation is embodied, is the field of tension set up by pairs of opposites, the energy generated by polarities held in fullness and strength. In a sense this field is the result of the organism's striving for a threefoldness which transcends logic – not one or the other but both together, where neither one diminishes the other but where each in its own realisation complements and strengthens the other. The third is then the culmination, the whole, which is enfolded in the parts. This particular field, or archetype, may thus be seen as a threefoldness which transcends, or resolves, dualism.

This tendency to threefolding takes place, though, within a larger field, which exhibits a fourfoldness. More than simply providing the energy through which change and movement occur – or through which life maintains its essential dynamic – this fourfolding embraces the creative impulse by means of which, and in service of which, an organism evolves and develops to realise its potential. This fourfold movement contains the elements of the creation, as known to antiquity – the archetypal elements out of which the world arises: fire, air, water, earth. Taken together these form the basic pattern by means of which an organism is enabled to realise itself. Where threefoldness provides pure energy, fourfoldness provides also direction and intent.

The cycle of creation follows a particular sequence in the movement from one element to the next. Yet it is dangerous to assume too much linearity in the way the pattern is formed. The sequence outlined below can be seen as an aid to understanding only; in reality each element is operating all the time – a healthy whole will be balanced. A balance not frozen but alive, so that at times certain of the elements may have to be emphasised, or strengthened, over others; certain elements may have to be compensated for; but all depends on where a particular organism or social situation is at a particular time, in relation to its own creativity. Let's explore what this may mean.

We begin with the element of fire. Fire generates warmth, and is the first movement in the creative process. The quality of fire is that of transformation. Fire is the gift, bestowed upon humankind, which enables us to transform one thing into another, and so enables creativity. Imagine, for a moment, a process, or an individual, or a social situation or organisation, which has ground to a halt, lost its creative energy, its motivation, its sense of direction. We say that such a process has grown cold, lost its spark. And indeed it has – such 'characterisations' are authentic observations, accurate ways of seeing. The process has no more energy, there is no more enthusiasm, no more drive, no more will, no more belief that to continue will really transform and improve anything. Such are all indications – enfolded in our very language and common metaphor – that the warmth and transformative power of fire is absent. Fire is the key to unlocking moribund and stuck processes.

Fire, as an element which transforms substance through the medium of warmth, is associated with gestation, that miraculous process where seed or egg or embryo is held for long periods of slow warming in order that they may germinate and incubate, gather the forces necessary to manifest. Such processes belong equally to the social organism, or to the development of an idea, or to the emergence of a particular phase in the life of an individual. Warmth has also to do with a sense of security, with the provision of security, enabling the risky leap into new venture, the letting go of past comfort for the launch into an unknown future. Warmth must be there at the beginning of the cycle of creation.

And not only because it enables security and a slow gestation to take place. The new can never be achieved without some letting go of the old. Birth and death, after all, are also polarities, and some form of balance is necessary. When something is born, something dies. New ideas, new endeavours, new situations take the place of those which have been. Nothing remains the same. There can be no growth without sacrifice, and fire is also the element of sacrifice. This is so in ritual, but also in nature. Fire burns off the excess, the old, the unnecessary. Fire itself cannot exist without the sacrifice of the substance which is burnt. And both through sacrifice and gestation is the seed of the new enabled to begin its journey along the cycle of creation.

The cycle continues with the element of air. From the warmth of gestation the seed stretches out into the surrounding cosmos as it grows. A certain buoyancy is required in a situation which may have become weighed down by its own inertia. Such weight must be taken off shoulders if the endeavour is to show initiative once more. Air is the element of lift, of weightlessness. It is also the element of light. Beneath the surface, buried under the weight of past endeavour grown stale, or slowly gestating within the protective darkness of the womb's warmth, lies a new idea, a new concept, a new form or principle or possibility – a new life. It must move out from beneath the surface into the open air above. It must move from the darkness into the light – once again, a balancing of the polarities is required. Darkness was essential for gestation; light is necessary for growth.

So the element of air stands for light – to shed light on, to gain a new idea, a new understanding, to become enlightened. And more: to enable the new to emerge, such ideas must take flight, develop wings and fly, as it were. Air surrounds and towers over the earth. Air is the element most closely associated with space, and with the concept of 'in between'; air is not filled with clutter, it has no specific form, it floats and meanders and drifts from place to place. It is expansive. Thus the second movement in the cycle of creation. For a social organism, this aspect of this cycle is vital – it is the place where new ideas, new behaviours, new forms, new ways of thinking, are dreamed. It is the place of experimentation, and of insight.

Air allows us to breathe, and allows a situation to breathe once more. Breath itself is the vehicle of spirit – we refer to the breath of life, to God's breath being the spirit which animates life. Our breathing provides access to the miracle of life – we breathe in oxygen, and breathe out carbon dioxide. Within the imperceptible space between inbreath and outbreath the world is transformed within us. Our world is created through our relationship with what is given.

The element of air stands as the breath of new life, the space for new ideas, emergence into the light. It stands as insight and enlightenment, without which the cycle of creation will grind to a close.

And so to water. When we think of water we think of movement, of flow and fluidity. Water is the element of process. It overcomes rigidity, not through head-on and brutal confrontation but through finding the path of least resistance. Nothing, no matter how unyielding, withstands the inexorable flow of water. Through its gentle determination to move on, and through its ability to find that path of least resistance, it is able to prevail over that which appears much stronger than itself. Precisely because it is not rigid, it is more powerful than rock or concrete. Water carries the strength of humility. There is no structure which will resist the flow of an organism's process. There is nothing, no matter how unyielding, which will resist the flow of an organism's process, if such process is flowing.

Water is the third element in the sequence because process and fluidity are what is needed when the organism has become stuck, paralysed, and rigid. Once the gestation period has passed, and insight has enabled new possibility to surface, then the rigidity of the old must be broken down, old patterns let go of, old ways and structures dissolved. The waters must find the cracks, enter and gently massage as they flow continually onward, and so widen the cracks until there are gaps big enough for the new to enter. Current forms and patterns and structures must dissolve into chaos so that new forms may emerge. Water enables chaos so that order may be broken down and reformed.

But water is not the element of process simply because it dissolves and enables fluidity once more. Water, as we learned in Chapter 3, always tries to form a sphere; this is why it is the element of circulation – it always tries to join what is divided and unite it in circulation. The state of a social organism which has become moribund and stuck is of a thing divided; the living links which unite the different parts into one living whole have been severed, or have shrivelled and died; the energy is no longer flowing. Such an organism is characterised by fragmentation and alienation of the various parts from each other, so that the whole is no longer pulsing, breathing. Water is able to unite the parts once more into a living whole.

Open as it is to forces raying in from outside, its fluidity is able to allow the outside environment to enter the organism once more. Social organisms are open systems which must interact with their environment to survive. One of the greatest dangers is a stagnant organism which, in its defensive rigidity, cuts itself off from its environment, and thinks of that as strength. It is not only weakness but the beginning of a death. And perhaps the deaths of others as well; such an organism is not only a danger to itself. Water, as the image of process and flow, enables vulnerability and humility in encouraging an organism to open itself up once more.

Yet too much openness is also unhelpful – it breeds confusion and chaos. Too much diversity leads to fragmentation. Thus, in the creative cycle, while the old must indeed be let go of, the organism cannot remain indefinitely in a state of open process. The new vision, the new approach, the new departure which has been stimulated into being through gestation, insight and movement must be brought into form if it is to endure. The old rigidities and structures may have been broken down, but now new structure must be formed. Once the past has been let go of, the new – which is as yet a mere seed or nascent concept – must become grounded in new practice. Spirit must become embodied in matter; or, as Goethe would have expressed it, pure being must manifest as material appearance.[1] Impulse, which gives us wings to fly, must be grounded, lest it leave us 'with our heads in the clouds'. Even as old structures and constructions must be rendered fluid, so too fluidity must be given form – another instance of the turning of polarities. So we come to the element of earth, the image of matter, of physicality, of structure.

Earth is hard, enduring. It is solid and formed, in a very literal sense, grounded. Images of foundation and bedrock emerge. One can build with earth, give form to concepts, enable ideas to manifest. Enable the nascent new to become something substantial, allow the creative process to materialise into something which can be seen, touched, produced, shared with others who have not been part of the process. For the social organism itself, this may manifest as a new product or service, new structures and ways of relating, new forms of accountability, perhaps, or of communication or decision-making. Particularly, it should manifest as new practices, such that the new is grounded in the way that the organism comes to behave. So the new is finally embedded, embodied in the organism, and the organism itself is new. It has re-created itself, moved beyond its previous high-water mark, evolved into something beyond itself. So does the cycle of creation build the new from the husk of the old.

Of course, the new will then once again become the old. As impulse and process are embodied in matter, so the physical begins to gain dominance; the effect and influence of spark, breath and flow become submerged within the preponderance of convention. The social organism begins to regard its current structures and procedures as its defining characteristics, not seeing that structure is always a product of past initiative, never of the future, and that procedure is hardened process.[2] Vision, ideal and intent give way to scepticism, compromise and the manual of rules. Risk gives way to defence; openness to an emphasis on boundaries. Once the new has become common practice it ceases to be the new; it becomes the old, defended unto death – and a new cycle must begin. Once the invisible has precipitated out into the visible, once spirit has fully entered into matter, then the visible and material will age and die, leaving a corpse, to be reinvigorated once more. Thus the turning of the seasons, thus the turning of pairs of opposites, harnessed within a fourfold cycle in order to enable life to move on, evolve and express itself in new and more complex order. So the fourfold field incorporates the four elements as images of its cyclical movement – warmth, light, movement and grounding.

There are social organisms which are immersed in the warmth of human interaction and nurture to the near exclusion of all else – many families, some communities. Pioneer organisations (organisations in the early stages of their existence) on the other hand, immerse themselves in warmth and light – they focus on the new possibility which they represent, and carry it through informality and loyalty; not for them the strictures of structure and regulation. Loosely formed social movements are similar in pattern – and it is often dangerous to the integrity of the social organism to insist on incorporating other elements before the real need for them has arisen. (By 'real' here I mean authentic within the organism's process, not assumed from without.)

There are organisations which focus on the fluidity of process – certain professional and service organisations, for example, need to be mobile and highly responsive to client need. Yet this may turn detrimental when given too much emphasis – as with the organisation which is highly expedient and devoid of value or principle, so that it will do anything to increase its slice of the pie. There are bureaucracies which are so focused on structure and procedure that they have become cold and inhuman places in which to work, devoid of participation and interaction and focused more on maintaining the internal status quo than on its service or practice or product, the needs of its clients, or its own strategic intent.

Development practitioners have to learn to read such social situations so that inconsistencies, contradictions, imminent imbalances, are revealed in good time, that organisms are helped to balance and to draw energy from their own dynamics; and to read their own dynamics in the future, in order to adjust themselves.

EXERCISE

Growth and Decay

We now combine some previous exercises – drawing and observation of plant metamorphosis, and polarities of sense observation and imagination – to further strengthen our understanding and ability to see life as developmental process.

We return to our experience of drawing and observing the metamorphosis of plants – if you have not engaged with such process thus far, this is a further invitation to do so.[3] Observe particularly two phases of plant development: in the first place, budding, sprouting and blossoming; and in the second, fading, withering, dying and decaying. In the first place, observe such processes carefully and with great attention to detail – drawing will help, as will listening to the different sounds made in blossoming or withering plants by the wind, or by an animal passing through them.

Notice how, in the process of growth, the leaf or flower or bud unfolds through relationship with the warmth of sun and the light and dampness of the air and with the movement of water and mineral, such that the smallest part of what you are observing grows in remarkable harmony with that which surrounds it, eventually with the whole plant and the environment as well. Notice how a whole is formed and enhanced out of inseparable and entirely interdependent parts.

Then pay attention to the process of decay, how the plant, the leaf or flower gradually dries and disintegrates, so that every smallest part of that leaf or flower separates out from other parts, begins to lose its connection until the living whole is no more, and only discrete pieces remain, to disintegrate further into smaller and smaller components, incapable of forming a whole. Notice how the same forces of sun and air and water and mineral now play a major part in accelerating this process of decay.

Once such observation has occurred, allow the sense impressions to play themselves out within your thought life. Withdraw from observation and allow the process of growth and decay to perform through your imagination. Follow what your eye had seen, but gradually allow your inner life to respond within the process of imaginative reconstruction of events. What are the feelings which arise within you as you experience the growth of the plant within your own imagination and soul; what are the feelings which arise within you when you experience the process of withering? Can you begin to experience, or understand, the inexorable necessity of such processes? Notice that the same forces which contributed to growth also contribute to decay. Allow to unfold within you a sense for the forces which live within the plant and which use the elements of warmth and air and water and mineral to unfold its process.

In this way, not only may a living picture of the process of development and metamorphosis emerge, but we begin to experience our own processes of development as an inner mirror of what we have experienced in the world outside.

We can then contemplate our own thinking to understand and strengthen it further. The pattern of growing and withering becomes directly experienced here. Notice how thoughts are generated – they flow, unfold, emerge out of the totality of thought-possibilities which we have at our disposal; and they maintain a living connection with the myriad thoughts which swim the seabed of our inner thinking life. So they grow and live – through the grace of their connection with the totality of our thinking. But if clarity of thought is to be attained, these new thoughts must begin to separate out, differentiate themselves from the others; they must become distinct through acquiring boundary and border. They must be capable of being lifted out and presented to others outside of oneself – thus the process of consciousness and articulation.

But even as we do so, the thoughts begin the process of withering and decaying. As they (necessarily) lose their connection to the whole, they become rigid and fragmented, no longer fructified by the forces which gave them life. Thus may inspiration metamorphose into plan; thus may mobile idea, imbued with warmth and immediacy, become abstract concept, eventually reduced to brief instruction or injunction. So may holistic understanding, which may generate enthusiasm, be reduced to the bullet point or technical diagram which fragments the whole into parts.

The entire process of growth and decay is necessary and inevitable. Learning to identify its movement, both within ourselves as well as in the world without, will help us find the necessary balance and timing that the development process may unfold freely.

10
On Becoming

*For nothing can be sole or whole
that has not first been rent!*

William Butler Yeats

We said that three major patterns or archetypal fields of force would be outlined before we move on to looking at the lesser patterns contained within them. We have dealt with two: threefolding and fourfolding. The third provides the deep rationale for these two and brings in the concept of oneness or unity.

The whole can be characterised as movement. But it is more than simply movement – it is *directed* movement. The whole is always in a process of becoming; and this process of becoming is a journey towards itself. It is always striving to become more *itself*, to become more fully realised. In the case of human organisms – individual as well as composite – which enjoy the privilege of self-consciousness, the journey towards wholeness encompasses the gradual development and enlarging of consciousness. And particularly *self*-consciousness – becoming whole implies becoming more aware of self, more conscious of who we are; it implies becoming transparent to ourselves, so that we come to *know ourselves*. This in turn means gaining freedom from the influence of unconscious (invisible) fields or forces. Thus do we gain a degree of authorship with respect to our own lives, rather than being helplessly (and hopelessly) the subjects of circumstance.

Jung once observed that that which has not been brought to consciousness appears outside as fate (while that which has, manifests as destiny). The degree to which we are able to take responsibility for our lives, the degree to which we are able to exercise freedom, depends directly on our level of consciousness. The journey of becoming is a journey towards such consciousness. This applies in equal measure to social organisms, which are created in our own image. Every such composite organism, whether small temporary group or organisation or institution or community, is in a process of development. This development trajectory shares the kind of patterning which we are talking of here with all other social organisms, and is simultaneously unique.

Every organism will experience periods of clarity, energy and enthusiasm and times of sleepiness, paralysis, stagnation. Will lose the clarity it once had; will struggle to find it once more. Will experience periods of renewal and (re)discovery. There will be times when structure, rule and regulation take precedence over human interaction and participation. Equally, every such organism will experience the enlivening vitality of human synergy and excitement. The process of development, which is the journey of becoming, the journey towards wholeness, is characterised by a rhythmic pulsing in which wakefulness is followed by a falling asleep is followed by a coming awake is followed by sleep. *We cannot come to greater consciousness in any other way.*[1]

Learning something is a revelation; it enables both the organism and the world with which it interacts to be lifted to a new level of existence. Learning something is a creative act in itself – it reveals new aspects of the world, and thus it enables the world to become something more than it was. We are the world's sense organ, and enable it to become more than it was. We do not create meaning – we enable meaning to emerge, and the more that emerges, the more the world becomes.

But once something has been learned, we come, over time, to take it for granted. And we no longer really *see* it; it becomes part of the world as we know it, unremarkable, uninspired, uninspiring. We see the world through it, and so we come to no longer really see the world at all. So does the world shrink, diminishing in response to diminishing consciousness. And the social organism itself, though it may have achieved much, has now gotten stuck, in a particular image of itself, in particular ways of being and doing, in particular relationships. The whole is no longer in movement; it is therefore in danger of dying. The natural world is not subject to such danger; but where self-consciousness exists, we become responsible for our own evolution, and can therefore lose the thread of our own story.

Consciousness, then, is about learning. Not about *having learned,* nor about having learned *something,* when these have become routine and accepted 'truths'. No, it occurs at the time of learning itself. It is the revelation that learning brings that enlarges the organism and its world; the *process of learning* itself is the path of becoming.

The organism is on a journey towards itself, towards consciousness, freedom and responsibility (that is, the 'ability to respond' to whatever comes towards it). Every social organism is different; every one brings something unique to the world, has something unique to contribute; the journey towards consciousness is the journey towards realising that unique contribution and thus becoming author of its own destiny. And it does this through the medium of learning.

A few words, though, about learning. We can learn new 'things', develop new skills, gain fresh understanding, through reflecting on our own experience or through incorporating outside input, perhaps through reading, perhaps through training programmes. In this sense learning is incremental – it adds to what we already know. And incremental learning can indeed enable a social organism to gain necessary competencies.

But this is only one way in which we can understand learning. The learning we refer to when talking of the journey towards greater consciousness of self is equally about learning ... not about 'things' or the 'world out there', but about our own way of being. Learning in this sense means becoming transparent to ourselves, so that we are able to recognise and work with our tendencies, biases, prejudices, mindsets, derived attitudes and habitual behaviours. (And, ironically perhaps, by doing this see the world beyond ourselves afresh.) It is in this sense that I refer to learning as the means by which organisms become more themselves, more whole.

But in this sense, we do not really refer to learning at all, but to *unlearning*. It is important to be clear about the distinction, because we can often be too glib in talking and writing about 'the learning organisation'. Learning is incremental. Unlearning, on the other hand, is making transparent all that has been learned up till that point, in order to choose the future afresh, in order to see the world anew. Unlearning is about making conscious all the unconscious habits and routines and ways of thinking and behaving and valuing which have gradually come to form the (largely hidden and invisible) group culture, in order to regain authorship and responsibility. Unlearning is about seeing how the organism's movements, form and content have been influenced by – and even trapped within – invisible archetypal fields, so that the organism can begin to act in freedom once more.

As an organism grows it assumes behaviours, ideas, values, structures which help it to become something of substance. But, as circumstances change, and as the organism itself changes, these ways of being and doing often become less than helpful. Sometimes even destructive, often constricting and eventually always limiting. They also become – the culture, the whole, becomes – unconscious. To render consciousness once more, the organism must unlearn much that it has come to take for granted. The development process of a social organism is partly, then, about learning, in the sense of gaining skills, knowledge and capacity; but it is equally about unlearning, in the sense of gaining freedom from past limitations and constraints. Unlearning enlarges and expands an organism, endowing it with more wisdom.

The forces arrayed against such processes of unlearning and development are immense. The organisation, and the individuals within it, fiercely resist change, and hold on with often hysterical tenacity to old ways of being, however uncomfortable and painful they may have become. Changing an organisation's culture is the most difficult task facing either the organisation itself or the social practitioner. It is long-term work, complex and daunting; yet it is the only work which will really have social benefit, in the long run, for it is about social organisms renewing themselves.

In this struggle, achieving major change is exhilarating and rewarding, but it should not be regarded as the only objective. The struggle itself, the struggle for a greater wholeness, the struggle to effect change, is the real reward, for this is the struggle for consciousness, for a viable and sustainable social order. It is in the process of learning – and now unlearning – that consciousness arises; once the organism has learned, it will begin to close down once more. Keeping that space open is the greatest good one can do, either as social organism or as social practitioner.

The organism which closes itself off from its environment will gradually lose energy and capacity, until it becomes either irrelevant or dictatorial. A social organism which cuts itself off becomes contracted, defensive, closed; it cuts itself off because it fears (rightly) that contact will necessitate change. The process of becoming whole demands that organisms keep themselves open and permeable; the process of becoming is a journey towards openness. Such openness will also assist social organisms not to turn on others in defensive reaction; it will minimise their projection of their own shortcomings onto others. A group which is consciously striving to achieve some transparency with respect to its own processes, is a group which is likely to contribute to, rather than detract from, the social good.

In moving to a new level of awareness and functioning, the organism needs to let go of old ways of being (unlearn) before it is able to gain the reassurance it requires by taking on of the new. The new cannot emerge unless some space is created; and clinging to the old does not allow such space to appear. Creating space, so that an organism may enlarge itself, means overcoming the tendency to excessive and frenetic activity which is the characteristic of many contracted and defensive social beings.

This dynamic between contraction and expansion inherent in the journey of becoming is fuelled by the continuous turning between having and losing, losing and having. We lose what we have, only to rediscover it later, but this time at a different, more conscious level. We find what we have lost, only to lose it once more so that we can awake again in the finding. Perhaps, in attempting to find a threefold way of resolving the dualism between these opposites, we may express it like this:

losing – **discovering** – having

The joy and reward for wakefulness lies in the centre. Enfolded within the polarity is the balance. In this instance, the concept of discovery is particularly apt. For the journey of becoming, the journey towards wholeness, is a journey of discovery, both about the world which surrounds the organism as well as about the organism itself. And such discovery, which is really fresh apprehension of meaning, continuously enlarges the world, for that is discovered which could never have existed before. As an organism develops in this way, there grows in the world a new reality which never existed before, and the world changes. If the organism strives to consciousness, then our world as a whole becomes more conscious.

We began this chapter by characterising this particular archetypal pattern as oneness, or unity, and it is this because the path of becoming is a journey towards oneself. As we become more conscious, we grow ever closer to, and more acutely aware of, the unique essential of our life; and as we grow closer, our lives are enlarged through our developing consciousness. The more we know of ourselves, the more we become one with ourselves, the more we are able to experience ourselves as participant-creators of our circumstances. A journey towards unity.

Regarding the individual, Jung described this process of becoming conscious as individuation. It may be difficult for the social practitioner to transfer the concept from the individual to more composite beings. But this, I think, bears thinking about: if you approach a social organism as a being, a whole, with a unique destiny and set of possibilities, and see your role – whatever your relationship to that organism may be – as a helper and guide towards its becoming more conscious and aware of itself and of its role in the world, then your way of seeing the organism will enable that entity to become something which it would never have become without such regard and attention. We cannot avoid responsibility, and we are ever commited, even when in spite of ourselves.

EXERCISE

Reading Oneself

Here follows an exercise which may enable you to explore aspects of your own path of becoming, to become more conscious of destiny and the inter-mingling of your own will with what is granted to you.

Think back over the last two or three years. Identify three people who have had a significant impact on your life during these years; the experience may have been painful and harsh or joyous and elevating, but whatever it was, your life would have been significantly poorer if not for these experiences. (The number three is not to be taken too literally; and these may have been new acquaintances or old connections with whom some significant interaction occurred during this time. These are indications or guidelines only.) Identify also three events of similar significance which have occurred over this time, which may or may not have involved interactions with people.

Having identified the people and events, ask yourself what they mean in your life. The assumption underlying this exercise is that there is meaning in what comes towards you from out of the future; that these are the questions and challenges you're needing to meet in yourself. As much as you create that future, so you are given that with which to create. This exercise is about beginning to make the process conscious. Jung said that that which is not made conscious appears outside as fate, while that which is made conscious appears as destiny, with which one interacts in the process of becoming.

So, look across these people and events which are very much part of your current process, all having taken place within the last while. Try to charac-terise the people and the events as beings or characters come to bring you messages – possibilities and challenges. Who are they and what do they mean? Try to separate the essential from the inessential. Is there any pattern which begins to emerge? What is it that they, or the pattern, are (is) asking of you now in the present? How can you best respond? Not only can you begin to get a vibrant and living picture of what your life is asking of you now, but you could begin to get some indication as to what is required of you into the future, to meet that destiny with clarity and integrity.

There are similar exercises which we can do with happenings that lie further back in the past, to begin to get some appreciation for the patterns interwoven into our destiny and our own patterns of response. We begin to gain an appreciation of how much is given that is of a value which might not have been seen at the time; how much grace is actually bestowed, as challenge and opportunity.

Look at the people who were significantly present for you at different stages in your life, and ask yourself questions of them similar to those posed above. Write character sketches of them; see them as archetypes you have met on your journey through life; try to penetrate through to the meaning which these encounters have had for you. See them as characters in a story of which you are the protagonist. Slowly you may begin to recognise how much of who you are has been formed through these significant others. Once again, separate the essential from the inessential; what forces for development were engaged through these encounters? Gradually, as we progress with this kind of exercise, we may experience a new depth in our existence, one which goes way beyond the transient personality which we had experienced as ourselves. The higher Self mentioned in 'Developing the Core' begins to emerge; we experience ourselves as simultaneously nullified without these decisive influences from others, and as an important and very large being who has been graced with these significant opportunities for development. Between the polarities of centre and periphery our process of becoming takes on depth and purpose.

Finally, we can consider those people who, right now, occupy our consciousness far more than others; those who are present to us continuously in a significant way, though we may not be currently interacting with them. They too are present in our consciousness for a reason; what are they asking of us, why are they occupying the inner place where we find ourselves? What do current configurations and patterns have to say about our path forward?

Exercises which entail reflection on aspects of one's life – of which there are a number contained in this book – may be done in solitude, but they may also be done in group settings. In such situations the group may be divided up into pairs, where one person undertakes the exercise while the other acts as a speaking partner, drawing out, challenging and supporting, asking questions, helping to take the process of reflection deeper, past possible resistances. In this sense the speaking partner is facilitating a developmental process (much as one would with a group) and there is good practice exposure available here. Once one person has completed their exploration the partners' roles will be exchanged. Once the pair are finished, the group may reconvene and learnings from the pairs be shared, although not the stories themselves, which will generally be too intimate and deep to share in such a setting.

Even within the pair, one should share only what one is able and willing to – these are development exercises rather than therapeutic processes. It often helps, though, even when there is not a bigger group present, to use a speaking partner in undertaking these explorations of self. We are helped in overcoming our natural resistances, in getting beyond thinking which may have become old and stale, and in finding a safe place in which to become vulnerable enough to go sufficiently deep that these exercises are worthwhile. It goes without saying that when pairs work together in this way, trust and confidentiality become key.

Note too that, with adaptation, these reflection exercises can also be facilitated with whole groupings, to enable such an organism to understand itself better. Such explorations, however, require high levels of competency on the part of the practitioner.

11
Dancing with Shadows

But what if I should discover that the very enemy himself is within me, that I am the enemy who must be loved – what then?

Carl Gustav Jung

Life, then, is not arbitrary and unconstrained anarchy; there are patterns which form it and give it coherence. Social organisms are clearly subject to these boundaries: the ebb and flow of natural archetypal patterns within which they take root and flower and fade. Three major patterns have been described; the interweaving of these patterns gives rise to specific tendencies – what may be termed minor patterns – some of which will be considered in the chapters which follow.

The first such tendency is that of the shadow, a concept developed by Jung as an important archetype within depth psychology. The shadow arises through having too much or too little of something, or through striving too much or too little towards something. When we direct our energies in too focused a fashion, a balance is lost; yet life strives towards balance. Thus the opposing polarity of our striving will manifest, and the more it is denied or avoided the greater its power will become, until the unconscious shadow aspect of our life becomes paramount, and we become trapped in its field.

The shadow is that part of us which we fail to see or know ... We are all born whole and hopefully will die whole. But somewhere early on our way we eat one of the wonderful fruits of the tree of knowledge, things separate into good and evil, and we begin the shadow-making process; we divide our lives. In the cultural process we sort out our God-given characteristics into those that are acceptable ... and those that have to be put away. This is wonderful and necessary, and there would be no civilised behaviour without this sorting out of good and evil. But the refused and unaccepted characteristics do not go away; they only collect in the dark corners of our personality. When they have hidden long enough, they take on a life of their own – the shadow life. The shadow is that which has not entered adequately into consciousness.[1]

The shadow is that which we specifically and consciously do not choose; it is that which arises when our attention is elsewhere. It is the counterbalance to that attention, the gap or void which is created and which cries out for attention, for a redressing of the imbalance. It is not an arbitrary manifestation but rather the unconscious twin of our conscious intention. And if the shadow is avoided or denied or rejected it gains in power until it becomes a potentially destructive force, a contradiction capable of fragmenting the organism, or at least of impairing its productivity and warping its energy.

But the shadow is not inherently destructive. It is the dark side to the light. The more light there is, the more dark there will be. The more energy we focus on a particular area, the more unseen forces gather in the shadows. Dark does not mean 'bad', or 'evil', or 'destructive'. Rather, the shadow lives as a potentially developmental nodal point around which the organism may turn. It is the obstacle which can shake us out of our complacency, precipitating the crisis, the need for which our conscious self may not yet be able to acknowledge. If taken seriously it can wake us up, cause us to take action before we fragment. In terms of the major patterns already described, it is the unregarded friend to our process of awakening and becoming, born out of the turning of polarities, and acting as the spur to shaking our fixed paradigms (grounded and stagnant) into facing a changed reality through the possibility of new seeding and growth. The shadow is the critical friend of lapsed consciousness.

Social organisms pour tremendous amounts of energy into making their intentions manifest, into realising their possibilities. Certain possibilities and certain attributes allow and encourage them to become what they are, and to realise their particular contribution in as focused a manner as possible. Yet even as they make this the subject of their striving, the dark consequences of their admirable intentions gather in the shadows, hampering productivity and warping impact. The way through does not lie in a switch to the opposite pole in the extremity of these contradictions; neither does it lie in looking the other way. It lies in coming to grips with the contradictions – staring them in the face, as it were – and consciously incorporating certain elements of the rejected extremities into the organism's reality. Such is the developmental and creative process of becoming.

For a more specific look at some of these dynamics, we turn to the story of a particular non-governmental organisation (NGO) which was experiencing certain problems at a particular stage of its life. We call the organisation Watchdog – not its real name. Watchdog is a civil society development organisation set up specifically as a voice with respect to environmental concerns. It is a regional organisation covering the concerns of eight countries.

At the time when these organisational problems were becoming particularly grave, Watchdog was a relatively small organisation – approximately 20 staff members. However, within environmental circles, within civil society and even government of that particular region, and within environmentally concerned civil society internationally, the organisation had an enviable reputation. It had become a powerful force. It had pioneered a perspective on environmental issues which highlighted their interconnectedness with the social and the economic – a perspective which has since become the norm. Information about the environment and social development was gleaned from people on the ground and interpreted in terms of Watchdog's holistic perspective. The results of this investigation and analysis were used to lobby those in positions of power, including governments and international consortiums and agencies. Watchdog had also assisted in winning respect – in the region and beyond – for the civil society organisation (CSO) as a vital and alternative institution, and had assisted in building the technical (environmental) capacity of CSOs and CSO networks.

Over the previous couple of years, however, Watchdog had begun to struggle with debilitating organisational conditions which threatened its very existence, and certainly its impact. In the first place, it was incredibly disorganised on many levels – its strategies had become incoherently diverse and fragmented, its financial resource base was rapidly diminishing and the prevailing atmosphere was of tension, conflict and strife amongst staff members, exacerbated by a growing lack of accountability between staff members and from the organisation towards its primary beneficiaries – those marginalised communities who had most to gain or lose with respect to environmental conditions and sustainable development.

Second, although it had indeed gained the respect of many, including its adversaries, it had of late – and to an increasing extent – lost such respect, ironically, amongst its friends and supporters. A growing number of these had come to view Watchdog as an opportunistic organisation, prepared to brook any contradiction in its search for power and status – or so they said. Being seen was becoming more important than contribution or substance; it was fast losing credibility, and credibility is really 'coin of the realm' so far as an advocacy organisation is concerned. Finally, while it was primarily an advocacy organisation, it did community development work amongst grassroots communities both for its own sake as well as to be able to take the concerns of the marginalised into the corridors of power – a unique strategic thrust. Yet of late it had lost the trust of the very communities it served, many of which were refusing to continue their relationship with Watchdog.

As a social practitioner called in to assist Watchdog, I struggled – together with a colleague[2] – to understand what was happening to this organisation which, despite its current situation, was staffed with high-calibre people dedicated to their cause. We could discover no subterfuge, no inauthenticity – yet the organisation was 'falling away', as it were, almost with a wasting disease; if authenticity and credibility were its life blood, then it was dangerously ill indeed. As we worked, we tried constantly to really 'see' the organisation, and to get Watchdog to 'see' itself. These processes fed each other, and gradually a picture emerged, constructed of a number of perspectives and insights which cohered into a real intuition of the organisational whole.

Here was an organisation with a very concentrated focus supported by numerous sources of inspiration. It believed that issues of environment could not be addressed in a piecemeal way – they had to be addressed holistically, because each aspect was connected to every other aspect; one could not separate and compartmentalise. This was a message which it tried to convey in the very manner of its advocacy work – it therefore tried to be everywhere at once, and to address itself to everything; it refused to be tied down, either to discrete strategies and programmes or to specific fora, locations or issues. It prided itself on its flexibility, mobility and responsiveness with respect to the way in which environmental issues were surfacing in the political and economic environment – it had developed an ability to carry the struggle with great cogency into whichever arena offered an opportunity for advancement at any particular time. To do this it purposefully travelled as light as possible, with respect to organisational 'baggage' – it kept structures and procedures down to a minimum, allowing staff great freedom and creativity in the execution of their tasks. It prided itself too on its direct contact with grassroots communities, and on the fact that its advocacy work was not simply backed by academic research but more importantly by the real needs and perspectives of real communities who were suffering the effects of environmental abuse, and who had also to reconcile their need for economic development with a sustainable environment.

Laudable intentions. Yet in their very focus they had unleashed their shadows which – unremarked in the nether regions of the organisational unconscious – had begun to play havoc with conscious intent. The aspiration to an holistic approach had rendered the organisation incoherent, fragmented and ad hoc. Its commitment to the systemic and interdependent nature of environmental issues had transmuted into a belief that every activity must have an impact somewhere, and thus the specific impact of different strategies had become diffuse and contentious – the 'holism' argument had degenerated into a lack of accountability and a kind of licence to do whatever people wanted to. At the same time, its emphasis on flexibility and mobility compounded the issue – while the point, for an advocacy organisation, makes great sense, it had become subverted into a desire for power, voice and status; the shadow manifested as manipulation, expedience and greed. Watchdog's lack of structures and procedures simply added to the confusion – it was not only the outside world which reacted badly, if unwittingly, to the organisational shadow, but the staff themselves saw their colleagues as unaccountable, opportunistic, and incoherent; so that conflict and tension had come to replace the original spirit of creative teamwork which had previously characterised Watchdog.

Finally, and perhaps most tragic of all, was the fate of Watchdog's involvement with community development work, undertaken in its attempt to bring the concerns and realities of 'common' people to the attention of the powerful. Watchdog's commitment to advocacy remained its focus, and because it had turned to community development work largely for the sake of advocacy rather than for the (direct) benefit of the communities themselves, it had to face the charge of using the very people it claimed to be acting for as a means to its own ends. The communities had already detected this aspect of Watchdog's shadow, even if the organisation was still blind to it, and were growing more and more resentful of the fact that their privation was, in a real sense, subsidising the global lifestyle of those who worked for this increasingly unconscious and defensive NGO.

In a few words, then, Watchdog's shadow can be characterised thus – that in its very attempts to become an alternative form of institution capable of a kind of guerrilla warfare conducted from the moral high ground, it had released forces which prevented it from institutionalising itself. Further, it had subverted the honour of guerrilla warfare into the ignobility of self-aggrandisement, and the moral high ground into a cesspool of intrigue. And in its activities with communities it laid itself open to the charge of being undevel-opmental, which meant that it was perpetuating precisely that which it was struggling against. Mobility and principle had degenerated into expedience and presumption.

The interweaving of the three major patterns is clearly evident in this story. The turning of polarities is very obvious – both in the way that the shadow manifests and in the palpable searching for a balance between opposing poles. The particular pattern formed by the cycle of creation is visible too. On the one hand, Watchdog is all air, and some fire – there is little of the self-reflective movement of water apparent in its brittle defensiveness (indeed, its emphasis on mobility has become a parody of water), and it has refused to ground itself to the point where its very sustainability is threatened. Looked at from a slightly different angle, its shadow has manifested as a grounding in spite of itself – the refusal to intentionally ground, to recognise the necessity of this archetypal pattern, has brought it to a grinding halt: paralysis and presumption are indeed the shadow of the earth element. As noted earlier, the pattern will be realised, one way or the other – our freedom lies in working with such archetypes, rather than in believing we are beyond such constraint.

Finally, if we can 'see' the organism (as whole) as a process of movement on its journey of becoming, then the meaning of these events, and of the manifestation of shadow, becomes clear. This story is just a phase in the organism's life, a critical phase in which the shadow has precipitated crisis in order to redress imbalance, to bring the organism to consciousness once more, to enable it to find a new level of path and being. As Watchdog became aware of its contra-dictions, its consciousness grew through the pain of its realisations and in the midst of its struggle to deny the need for change – and it recognised how stuck it had become, and what it needed to do to continue its journey. Not to deny its original values, but to incorporate its shadow – build a modicum of structure and procedure, either work properly with communities or not at all (both of which imply the renewal of strategy and methodology), develop the holistic approach into a coherent and accountable methodology, and develop the respect for people (both inside and outside the organisation) that it has always accorded to the environment. In this fashion, genuinely build an alternative form of institution – which was its original intent – by learning from its experience, and unlearning those habits and modes of behaviour which have proved debilitating. And so, gradually, take another step in becoming the social organism which it has the power to be.

EXERCISE

Exploring Shadows

Here follows another self-reflection exercise.[3]

Identify a problem in your life which is intransigent, an aspect which you have wished to changed but without success. A place where you are stuck. A recurring difficulty or issue which you are not happy about, and which you have tried repeatedly to change, but to no avail. A measure of frustration has crept in; it matters not at this stage whether you attribute the cause of this difficulty to yourself or to persons or circumstances 'out there'.

If you have a speaking partner, describe the issue generally as an aspect of your life, but then hone in to a specific incident which serves as an instance of this issue. Describe the incident in great depth, taking care to make it as vivid, as pictorial as possible; your speaking partner should be able to experience the incident as if he or she was inside it at the time. Describe the circumstances surrounding it, what happened at the time, your frustrations and unsuccessful attempts to shift the dynamic. Describe what you did, in detail, and what the other person or people did, also in detail. (Note that with all of these exercises it is in the detail, in the authenticity and vibrancy of the story, that the whole – the meaning – is observed. Vagueness or generality will disable the process of reflection itself.)

Making a jump (perhaps) to accepting responsibility for this situation, rather than placing the blame outside of yourself (even if at this stage just hypothetically), ask yourself this question: Given that you did not succeed in changing the pattern, what did you lose by maintaining the situation as it was, and what did you gain? This last question may be particularly difficult to face, because you might never have thought of yourself as gaining something from maintaining a difficult dynamic which you imagined you were desperate to change. You are indeed desperate to change, but you are both gaining and losing by maintaining the situation, and on an unconscious level you are gaining something which you do not want to give up. Between losing and gaining, that unconscious aspect of yourself (which is gaining) is constantly winning the battle.

Having come this far, try now to characterise that (unconscious and hitherto unacknowledged) aspect of yourself which is gaining. Describe its essence, describe it as a being, a character. That being is the shadow we are in search of. Perhaps not the only shadow, perhaps only an aspect of shadow, but the unacknowledged twin of your conscious striving, nevertheless.

Having identified it, having brought it out into the light, try to identify the first small steps that you may take to really transform the dynamic, now that you can see your own part, responsibility, in the situation. But respect too the tremendous energy which such shadow may be providing you, and think about ways in which you can harness that energy in consciousness. Simply having made the shadow conscious will already begin the process of transformation.

12
Paradoxes of Power

Lost in the solitude of his immense power, he began to lose direction

Gabriel Garcia Marquez

The journey of becoming is the primary mover behind change in human systems; development is a process of increasing consciousness. A very forceful dynamic which can disable or enhance such development is that of power. As with the shadow, power emerges from the interweaving of the three primary patterns as an irreducible tendency which has its own, unique form – power manifests as paradox. In this chapter we deal with the pattern of power by using a number of different entry points to build a multilayered perspective. Of all patterns, power is the most complex and obtuse.

As an initial foray into the subject, we take the story of a network of 13 Civil Society Organisations, spread over eight African countries, involved in a particular sector of development work. Some years ago this network conducted an organisational assessment of the strengths and needs of its members. The results of this assessment initially confused and intrigued the membership. On a simple, material level the results of the survey were clear. Some organisations were more effective than others, some were strategically more coherent, some were better staffed, some were better funded, some worked within more enabling environments than others; some were stronger, others were more needy. But it proved difficult to correlate the strengths or weaknesses of particular organisations with their outer circumstances; indeed, there was a large degree of divergence between outer conditions and organisational capacity. Simply stated, lack of organisational capacity did not seem to correlate easily with debilitating outer circumstances; and change in material conditions did not seem to be the answer to organisations' problems, for they had not caused the problems, nor had they been a solution for the more capacitated organisations.

Consequently, we had to search for causes beyond the material. It transpired that it was not so much conditions or circumstances which were a cause of organisational capacity or the lack of it, so much as the organisation's *attitude* to these factors. Those organisations which complained about their material lacks, which attributed their problems to the world 'out there', lacked the ability to counter these problems; while those organisations which were more self-critical, attributing their problems to their own lack of strategic coherence or management competence or focused vision or evaluative (self-reflective) stance, manifested greater organisational capacity as well as the ability to overcome or compensate for outer constraints. The correlation between organisational effectiveness and organisational condition seems to lie in organisational culture, in the organism's prevailing attitude, rather than in circumstance (though of course circumstance remains an important and influential factor). The key to an organism's power lies in its attitude, and the key attitude appears to be responsibility for one's own circumstances.

This finding is not new – it has been apparent for some time that organisational capacity, and the source of power, lies in the invisible and intangible aspects of organisation: in attitude, culture, conceptual understanding, strategic coherence, leadership.[1] This is, indeed, precisely why social practitioners are called on to be artists of the invisible. But this does raise a question about power. For if the source of organisational power is to be found internally – in organisational attitude rather than in outer circumstance – then whence come the differences between organisms in the first place? If attitude is both source and limitation of power, what gives rise to attitude?

Fritz Glasl, an eminent organisation development consultant and writer who has specialised in the area of conflict resolution, offers this illuminating perspective on power:

> Power is the potential of Party A ('the powerful') in a given social structure relationship to impose her expectations on Party B ('the powerless') in such a way, that Party B sees less chance not to comply with the expectation of Party A.[2]

We note a number of aspects to this perspective. Structure, the imposed situation, is a determining factor. This would seem to imply that the powerless are indeed such, with no possibility of alternative. However, the second part of the statement casts a somewhat different light. Here, the way in which Party B sees the situation becomes a determining factor. It operates much like a game of cards. The hand we are dealt is given; our choice lies in the way we play the hand. The way in which we play it is largely influenced by our expectations of the situation, which, in the case of the powerless, is largely governed by the framework set by the powerful. But the 'play' of Party B is not determined by Party A; it is determined by Party B's expectations. The latter are largely governed by Party A, but the way through the impasse is not to expect any change in Party A's behaviour, or even to demand such, but rather for Party B to change his or her expectation of the play, and own behaviour. To break the dynamic, as it were.

It is the attitude of the powerless which largely maintains impotence, or which at least constrains alternative. This attitude is generated through past experience of what may happen. Note that the statement refers to power as 'potential'. There is a crucial angle here. The expectation of power is greatest just before it is actually exercised; it is much greater in potential. Once the worst has happened, the tension has been drawn out. There is, then, a collusion between powerful and powerless; the powerful might not be able to control, were it not for the subservient and compliant attitude of the powerless – generated by fear of the potential. The powerful can have little fundamental control over those whose will has not been broken, or over those in whom the will to power has been woken.

Thus we arrive at one of the essential paradoxes of power. The expectations of the powerless are determined both by circumstance and by the imposed expectations of the powerful, but any change will be effected by a change in *attitude* on the part of the powerless, not by circumstance, nor by the powerful. Thus, to turn back to the story with which we began, attitudes which constrain effective behaviour, which limit potential and capacity, may well be due to imbibed pictures and experiences from some place beyond ourselves, and even beyond a particular social organism, but any search for a shift in these attitudes will have to take place within ourselves, within the organism. We can demand nothing if not our own resolution; we can assume nothing if not our own response.

This is the first paradox – that the powerless collude with the powerful to maintain the hegemony of the powerful, and this collaboration is an unconscious one. The possibility of change lies largely with the powerless. The move towards a shift in attitude on the part of the powerless is both a move towards consciousness and a result of having broken through unconscious barriers. The embedded expectations which prevent the powerless from exercising their power thrive only through lack of consciousness; indeed, they are a manifestation of it.

To approach the issue from another angle, we can ask what happens when the powerless do change their attitude, and begin to manifest power through taking responsibility. When power has long been exercised in the hands of some, it tends to concentrate in the centre of the social field (a point we return to shortly) as a dominant paradigm. What little alternative voice is left – or beginning to grow – manifests its power from the periphery. The borderline is ever a place of last remaining freedoms, a wild place beyond the reach of the prevailing norm. Yet such is the paradox here that the very exercise of such power risks its demise. As the peripheral voices gain credibility they get drawn into the centre, where freedom of expression, even of thought, is usurped by the dominant paradigm. Such at least is the danger when the dynamic of power begins to turn.

The organisation (CDRA – Community Development Resource Association) with which I have worked for many years is a case in point, and an example which is close to me. The CDRA began in 1987 as an NGO dedicated to providing developmental consultancy services to other non-profit organisations in a very localised area of the south. It started out as an experimental idea, providing facilitation services in organisation development, while at the same time building its own understanding of, and expertise in, organisation development consultancy. At that time, within the development sector, the discipline of organisation development was largely unknown and unused. From the beginning, CDRA dedicated itself to providing an effective service and to perfecting its own expertise in delivering that service. It avoided making grandiose claims either for itself or for the discipline of organisation development; it steered clear of rhetoric, of forums and networks, of what it perceived to be the dangers of too much loose talk; it maintained a low profile, preferring to work in the background, with small local organisations, and prove value through results.

As the years passed the world changed. Organisation development consultancy became a recognised service amongst the organisms of civil society, and such organisms – indeed, civil society itself – gained credibility and profile. Development consultants, and development practice, are a growing industry, albeit a contested and ambiguous one. The CDRA, having pioneered a particular form of development practice and consultancy, and having concentrated on developing its practice and its learnings from that practice, and having gradually – through its work and its writing – gained privileged access to a large spread of the development sector internationally, is now faced with the dilemma of growing power, revealing the dangers of coming in from the margins where it has lived for all these years.

The development (or aid) sector, as a whole, struggles to find its way, to prove its worth, in the increasingly complex and and morally contentious wake generated by the twin triumphalisms of globalisation and capitalism. Meanwhile the CDRA, while remaining a very small and concise southern organisation, has been able to increase its influence and impact immeasurably. Increasingly too, it feels that it has things of urgent importance to say to the sector, ideas and practices and perspectives which have emerged from its practice in the sector over the years. It is now able to command a certain respect. And this ability confers in turn a certain power. A power which emanates from out of, as a part of, and on behalf of, the margins. Yet even as it dares to own this power, it is seized with a foreboding sense of the dangers inherent in taking such possession. The paradox of such power is such that its use runs the risk of drawing the organism into the centre, and away from its source of legitimacy, value and relevance; even from the source of its truth. And this in two senses.

First, the source of CDRA's power is its relentless and rigorous focus on practice. Yet as it increases its influencing and advocacy role, it is forced to engage in strategies which are a long way from the responsive provision of consultancy services for individual clients. The higher its profile becomes, the more it is drawn into the 'development circuit', and the less it practises; and the less it practises, the greater the danger that genuine power will drain out of the organism's profile until it begins to caricature itself.

Second, as a small and southern CSO, CDRA is a marginal, peripheral entity; an alternative at best, a chimera at worst. But then, to remain so, what is the point? It is the paradigms, the policies and the practices at the centre which dominate, and without change in this arena CDRA's work on the margins becomes ephemeral, an everlasting 'attempting', with little solace. Yet as CDRA engages more with the centre, less time is available for working directly with its primary constituency; in which case, whose agenda has hegemony? And inevitably one has to adapt one's discourse to be heard, and who, ultimately, is adapting to whom?

The second paradox of power lies in this: that for the powerless to gain power, non-engagement is not an option, while engagement risks reversing the very power gained. In the case of CDRA, there can be no avoidance of the dilemma; there is no way around it. Both sides of the dilemma must be held; a living, shifting balance maintained between the opposites of this particular polarity; there is, once again, no substitute for consciousness, for the balance will only be maintained through such consciousness. Indeed, the tension exists in order to maintain a level of awakeness – any lapsing into one or other polarity will diminish the power gained. Every social organism which attempts to wrest power for itself or others runs the risk of being co-opted into the dominant paradigm. There can be no certainty, no security, no one answer for the developing organism. The organism can only cultivate right attitude, consciousness and respect.

A third angle into the pattern of power takes its lead from the old truism that power corrupts. One aspect of this corruption appears to be a lapsing into unconsciousness, a 'falling asleep' into a dominant paradigm which usurps all others and which gathers to itself a certain arrogance, not allowing the powerful to see beyond their own truths and their own victories. Excessive power leads to presumption, and a casting out of alternative perspectives. What is not so obvious, though, is that such corruption, the abuse of privilege, does not only work to the detriment of those over whom power is exercised. It works as much to undermine and to disable that power itself. Thus is the third paradox of power revealed, a paradox which can be particularly debilitating for the organism in its process of development.

My work with development CSOs particularly, but also with both public and private service bureaucracies, reveals the same story which plays itself out incessantly as variations on this central theme. Those workers actually dealing, interfacing, with the organisation's clients, or target group, generally exist on the margins of the organisation – in the field, as it were. As such, they are low down on the organisational hierarchy; they often do not take part in policy or strategic decision-making; they are marginalised, as is their work, though it be the core purpose for which the organism was created. Power is concentrated in the centre, with those who literally stalk the 'corridors of power'; policy and strategy is decided here, amongst those who are office based, with the time and mandate to concern themselves with the organisation and its fortunes, rather than with the clients whom the organisation is serving. Those at the centre are usually the highly paid, the specialists, the analysts and theoreticians and tacticians, the educated; while those at the periphery are often the inarticulate, without sufficient breadth of vision, with their 'noses to the grindstone'. Yet, when this goes on long enough, the polarity begins to turn, partly because those on the periphery are not removed from the field, and begin to know more about the impact and effectiveness of the services offered, and the strategies employed, than do those at the centre who have the power.

Alternative viewpoints bring uncertainty, ambiguity, and often herald the need for change, which is the last thing that those at the centre want. They often simply cannot see the value of the information and viewpoints being generated by those at the periphery. They remain convinced of their truths, where in reality such truths have become merely the echo of their own organisational image. Thus do the powerful become trapped within the hegemony of their own discourse. For the centre's weakness lies in this: that it becomes unable to see beyond itself, unable to adapt, and some form of (often very painful) death is required before change and development can take place. Either such breaking must occur, or the centre clings ever more strongly to its discourse and practice, and a souring takes place, a diminishing of potential. Instead of expanding, instead of pursuing its journey of becoming, the organism begins to contract, and wither.

When the centre becomes too powerful, it rigidifies, lapses into anonymity, and a faceless, intractable tyranny results which no longer has any power to rectify itself, eventually even in the face of its own increasing awareness. The power at the centre loses control over its own development; it becomes disempowered through the rigidity to which it adheres. A closing off occurs, a cessation of movement. Regeneration takes place on the periphery, where the ragged and chaotic margins allow new elements to enter and combine with the old. The centre defends itself against such incursions. But, to paraphrase Bernard Lievegoed, the new is always met on the boundary of the shattered soul. The centre seeks order at all costs, yet without a cracking of the rigid shell, without the entry of chaos, nothing new can emerge.[3]

This third paradox reveals once again, as do the other two, that power is fundamentally a developmental phenomenon. Prevailing power derives from past victory, and thereby risks becoming trapped in its own grandiosity, which will eventually cause it to wane through the weight of its own assumptions. While the periphery holds the potential of the future, a waxing power to be attained through exercising the freedom to move, which is denied those at the centre. Real power lies in the process of development, of becoming. Clearly, it also lies in the process of awakening, of continuously discovering, mentioned in Chapter 10. For it is precisely at that moment when you know that you have it, that you begin to lose it. And only by losing it can you find it again.

As clearly as the pattern of becoming is manifest in the dynamic of power, so is the cycle of creation, for the movement from the seed of the new, germinating in warmth, through to the rigidity of the grounding process once it has become stagnant, is replayed here, and also mirrored in the movement of power from the centre to the periphery and back again. As the one pole rigidifies or fragments, so the other is able to entertain the seed which will renew the organism and enable it to continue its journey.

In a sense, the development process itself is a play upon the pattern of power, which is the movement from unconscious to conscious, from the unempowered margins of ourselves and the social organisms we create to the self-confident centre; and back again, in a spiralling and recurring motion which turns around on itself in its unfolding journey. The constant turning of polarities into their opposite, always in search of consciousness.

EXERCISE

Fingering Power

Here follows a way to get beyond what you may have read and thought about power, in order to access and learn from your own experience. This is a simple exercise, described below for groups, though of course various configurations are possible.[4]

Break a group up into two smaller groups of equal numbers – call them Groups A and B. Each person in Group A is asked to recall a situation(s) in which they felt very powerful. Each person in Group B is asked to recall a situation(s) in which they felt powerless. In each group, pairs are formed, and these pairs work on their own for a while, sharing their stories.

The pairs from Group A, as they tell their stories to each other, are asked to focus on a number of areas. Within a vivid and well told story, they should outline clearly what they did in the situation, and what the other person (the powerless or less powerful) did, as the events unfolded. Then to try and answer – with the help of the speaking partner – these questions: What made it easy for them, in the behaviour of the powerless, to go on with what they were doing? What did they have to access, and what did they have to close themselves off to, to go on? Are there any patterns which emerge from the stories?

The pairs from Group B are asked to do the same, but with slightly different questions. Having explored what the more powerful party did, and what they did, they can explore: Whatever the powerful party did, what did it do to you, as the powerless? What may your share have been, in what happened? What did you have to close yourself off to, and why? Are there any patterns which emerge from the stories?

At this point the groups are mixed, so that in the ensuing small group discussion there are people who recalled times of being powerful and others who recalled times of being powerless. These new groups now share the learnings, insights and patterns which may have emerged in their previous discussions, and compare them. In light of these discussions they may also want to discuss the ways in which facilitation might help, what the facilitator may bring to the situation, and so on.

These conversations come back to the larger group, which may then move in various directions, one of which could be to look at case studies of power dynamics within social organisms with which they work, to see whether this exercise may have yielded fresh insight into what is actually happening in such situations, and what kind of facilitation may be needed to help the situation resolve itself, or at least become more aware of how it might go forward.

13
The Pattern of Reversal

Every child is an artist. The problem is how to remain an artist once he grows up.

Pablo Picasso

We have been speaking of patterns. These patterns are laws of the spirit, the manner in which the spirit enables life to reveal itself, to develop. Without matter, spirit remains unrealised, an impulse only. Without spirit, matter remains formless and inanimate, an indeterminate mound of rubble.

To understand these patterns of the spirit it helps to reverse our usual ways of looking at the world – or perhaps, reverse the expectations which we have come to have of our world. It is this reversal which proves so daunting a task when trying to master our lives, for we have become accustomed to thinking of life as simply another 'thing' to manage, direct and control. But it is not another thing, and the laws which govern life are not of the same order as those which govern matter. To be fit to contribute to the life of a social being, an entirely different stance is available – to allow, enable, receive and respond – and so encourage the organism to assume such stance as well.

We have seen that the principles of logic derive from our experience with solid bodies. Thus, 'one thing is always itself and never anything else', 'something cannot simultaneously be itself and not itself', and so on.[1] Patently this is not true of that which we have been speaking of throughout these pages. When we look at the concept of the shadow, we recognise immediately that something can simultaneously be both itself and not itself; when we look at paradoxes of power it becomes abundantly clear that something can be both itself and something entirely other at the same time. The patterns of spirit reverse the logic of solid bodies, and the laws of matter.

We have just been considering some of these reversals in the previous chapter on power. We noted there that precisely at the moment when you know you have power do you begin to lose it, and that it is only by losing it that you can find it again. We noted that power grows through consciousness, but that power may grow to the point where it dulls consciousness, and thereby becomes the cause of its own demise; that power tends to concentrate in the centre, but that paradoxically at a certain point the 'unempowered' margins have more potential for power than the so-called powerful themselves; that the remedy for powerlessness lies with the powerless, not with expectations of the powerful.

This reversal is most clearly seen in our work on the shadow, where excessive focus or emphasis on anything will release its opposite. We have seen it in our exploration of the process of becoming, where – as Yeats put it – nothing can be whole that has not first been rent; we have to lose in order to find, we have to unlearn, shed, cast off so that we may grow and enlarge ourselves. It is clear also from the cycle of creation, in which the warmth of impulse is eventually but inevitably followed by the rigidity of rule and regulation; and in which such rigidity, built to contain and protect a new way of doing things, has to be shattered and broken in order to allow anything new to emerge, because it prohibits as much as it protects.

Our work on the transformation from twofold to threefold thinking continues the picture – we expect opposites to contradict each other, to void each other, not to complement, sustain and build each other. We would expect to find a balance between opposites through compromise, yet we are asked rather to increase the power of each opposite so that such balance may be found through the heightened tension between them. And that holding such tension is not aided by a fraught and tense disposition, but rather by a calm and tranquil centredness!

Spiritual laws reverse those of matter. They seem, to our logical mind, like contradictions. For instance, each part contains and forms the whole, yet the whole is not the sum of these parts, and the parts do not emerge from the whole – the whole and the parts co-exist, and form each other. Similarly, the more each of us understands the world, the more the world grows and the more there is to be understood – so we try to understand a world which we are ourselves creating as we go; the more we know, the more mysterious life becomes. Every move towards consciousness increases the sphere of the unconscious – deepens and enriches it; conscious and unconscious thrive on each other, and our world thrives on both. It is not either/or – the way of the spirit is both/and.

We lose our life in order that we may find it. We meet crisis and pain so that we may be jolted out of sleep into wakefulness; we may be grateful for those very events which seem most hurtful and incomprehensible at the time. When we hate, when we cannot forgive, we do no one more of a disservice than ourselves, for we tie ourselves and our energies to the very object of which we would most wish to be free. And we become ourselves that which we most despise. On the other hand, when we are most generous, we are most likely to be rewarded; the act of giving displays an openness which enables the world to return the gift a thousandfold. At the same time, an inability to receive gifts and offerings with grace will hinder the act of giving – for to receive with grace the offerings of others is itself a gift. In order to be trusted we have to trust – the only way to heal that feeling of betrayal is to offer trust, rather than to wait upon proof of the other's trustworthiness. The way of the spirit reverses normal expectations.

We cannot move on unless we are prepared to let go; and we cannot let go unless we have something to let go of, unless we have found something of value. We are not being asked to accept whatever comes (because life is full of contradiction anyway) but to meet whatever comes with a discernment built out of a deep and respectful understanding, and to attempt to respond appropriately.

The laws of the spirit are characterised by the approach of 'both/and', just as the laws of matter are characterised by an attitude of 'either/or'. To meet the archetypal fields of which we have spoken with courage and commitment demands both experience and mature reflection on such experience. For the prospect is daunting – it expects nothing less than full and devoted attention. Anything less will leave us precisely where we are – unable to cope with the testing nature of social life, and dominated by social trends and social organisms which are fragmented, untrustworthy and dangerously capricious.

I raise these considerations not merely to allow the weight of them to rest in the hand, as it were, so that their authenticity may be gauged, but to introduce a further archetypal pattern. It is when a system is enabled to unfold according to the pattern of reversal – which goes against the grain of all the training that we have acquired over the years – that it is enabled to reach its full potential.

A clear instance may be read through an aspect of organisation development theory which has assumed the status of 'lore'.[2] Organisations are thought to go through three characteristic phases as they develop (often interspersed with crises which herald the imperative of transition). Phase one is the pioneer phase, marking the beginning of the organisation, when charisma, creativity, impulse and inspiration collaborate to allow an idea to manifest in organisational form. This phase is characterised by warmth, informality and familial relationships, with the will-carriers leading the organisation through personality and innovation. The second phase (differentiated or scientific management phase) emerges when demands arise for formalisation and standardisation, when the organism becomes complex and complicated enough to require differentiation into departments, organs and specialisations, and when employees begin to regard their work as a job with boundaries rather than as a calling without constraint. This phase must begin to transform when the institution becomes too heavily bureaucratic, too rigid and inflexible, driven by its own internal logic rather than the demands of those it serves. When rule and regulation replace responsiveness, when strategic considerations centre around the needs of the organisation rather than the needs of its constituency, when employees lose a sense for the meaning in what they are doing, when the streamlined and efficient becomes the bloated and irrelevant.

Such renewal comes in the transition to the third phase, often called the integrated phase, when the positive aspects of the first phase and the positive aspects of the second phase are combined in an organisation which is structured and formal but which also engages with creativity and flexibility, where routines are followed but in such a way as to encourage mobility and responsiveness, where the meaning of the work and the vision of the organisation suffuse the various pieces of the organisational puzzle, and where individual responsibility is encouraged as an exercise in team accountability, rather than as opposition. It is recognised that such an organisational culture is difficult to maintain and that the organism will often slip back into either a first phase or second phase dominance; vigilance is called for.

On the face of it, then, we could regard the above phenomenon, in its resolution of opposites, as another instance of the necessity to move beyond dualistic to threefold thinking; there is then no new archetype or pattern of reversal revealed in this description. But under the surface there may be something more profound going on. The 'first phase' is really the pure outpouring of spirit, while the 'second phase' entails the fixing of such spiritual impulse into matter. The first phase is the realm of human agency, the realm of freedom and hope, of a striving towards the future, of choice and autonomy; while the second phase is the realm of structure, protection and constraint, of support and maintenance, of consolidation and affirmation of past impulse.

We cannot create a social organism with that which is emphasised in the second phase any more than we can create water from hydrogen and oxygen, or any more than we can create plant substance through combining carbon and water. It is not given to us to do these things; it is done by forces which sustain us but which we cannot control and do not create. So-called second-phase elements are a material manifestation of the spiritual and living impulse which is somehow – in some kind of alchemical process – released as the formation of a social organism, through the energy of the so-called first phase.

Put another way, the second phase may be something to be *achieved*, through our attempts to structure, regulate and control, while the first phase may be something to be *received*, a form of grace perhaps, something which requires openness, and trust in forces beyond ourselves. Of course the pioneer phase requires hard work of its leaders – but such impulse is the spirit bursting through, rather than the painstaking construction work which is required of the second phase, when brick must be placed on brick to ensure that the edifice stands. In the first instance, it is warmth and inspiration which carry the impulse; in the second, it is either the grinding labour of routine and repetition, or the long-term commitment of nurture and support.

The pattern of reversal signals that, for an organism to continually re-create itself, to develop,[3] it helps to remain open to impulses which come towards it from out of the future, as gift, rather than constantly and fretfully attempt to do its utmost to control and direct itself and its surroundings. To hold itself lightly, instead of tightly. Open and vulnerable, rather than closed and defended. The wholeness, health and value of a social organism is not only up to us, but up to that which we allow to work through us. The reversal consists in this: that the organism permits rather than constructs, trusts rather than defends – does the very opposite of what knee-jerk reaction would expect. So does spirit evolve; else we will ensure, as we are already, that beings emerge which are antagonistic to the flow of life itself, and we will come to serve that which we created to serve us. For the more we try to control, the more out of control we become. As Vaclav Havel put it: 'It is strange but ultimately quite logical: as soon as man began considering himself the source of the highest meaning in the world and the measure of everything, the world began to lose its human dimension, and man began to lose control of it.'[4]

The new sciences present a very similar picture with respect to all living systems. It is understood that living systems are self-organising – in other words, that they are *organisationally* autonomous. This does not mean that they are closed to their surroundings; rather that their ability to self-organise and re-create themselves is not imposed by those surroundings but is established by the system itself. At the same time, with respect to the outside environment, the *structure* of a living system has come to be regarded as dissipative – the system is *structurally open* so that matter can continually flow through it.[5] A dissipative structure does not resemble a building so much as a standing wave, which is formed, for example, behind the trunk of a tree submerged in flowing water. The system maintains a stable form, but one which is open to inflow, outflow and continuous modification.

Living systems are a mix of order and chaos, and thrive on the edge of chaos as here new order is allowed to emerge. Living systems will organise and re-create themselves from out of their own impulse so long as there is a continuous flow of matter through them – so long as they are chaotic enough to ensure that their boundaries are not closed to new 'information'.

If a social organism is a living system, then the message is clear: it must not be held too tightly, it must not be structured too rigidly nor managed too closely, rather it must be *allowed* to organise itself by keeping it open to new impulses and information. By *enabling* it to be guided by its values and vision, by its understanding and artic- ulation of its task, instead of insisting too much on rule and regulation, the organism will maintain vitality. It is not that the first phase must dominate the second, but that the third phase, the whole and healthy composite social being, is a *flow*, a continuous *process* of re-creation; it is not a fixed and ordered *thing*.

This pattern of reversal, which implies that less is more, is not only difficult to work with, but is in many ways unnerving. It is unnerving to think that we must do less in order to achieve more.

Yet, it is not so much a not doing, as much as a different way of being. The Tao states: 'Clay is shaped to make vessels; but the contained space is what is useful. Matter is therefore only of use to mark the limits of the Space which is the thing of real value.'[6] And

Aleister Crowley comments: 'Matter is like the lines bounding a plane. The plane is the real thing, the lines infinitely small in comparison, and serving only to define it.'[7] Which at least provides us with a way of seeing differently. As social development practitioners, it means that we work with the culture of the organism, its purpose and vision, its values and its understanding of itself and its world; we must work with the spirit which animates. The organism must be in touch with itself, and with the impulses which move it. If it is such, the rest will follow – it will organise itself, and accountability and responsibility will emerge in new and appropriate forms which will allow new ways of social organisation to emerge.

Practically, this means that perhaps our greatest contribution to a social system or situation may be to create space, and to allow that grouping to *be*, and not only to do. Continuous conversation becomes an imperative. Space to reflect, to share, to reveal preferences and tendencies, to encourage trust and personal warmth, to incubate new ideas and creative possibilities – all this encouraged and supported and nurtured, not within the context of timeframes and deadlines, not with a constant question of the immediate productivity of the session hanging like a vulture over the group, but in genuine freedom. Such space reverses every impulse towards 'good' organisation, which emphasises bottom lines, direct and immediate pay-off, the necessity of product over the luxury of process. Investments must demonstrate hasty returns. And as we concentrate so, the social fabric withers within itself. If we insist on bottom lines then this is what we will receive – lines (matter) rather than space (spirit), and the bottom rather than the heights.

Miha Pogacnik, a violinist who uses music in his work developing intercultural relations and social organisms, puts it succinctly: 'Mastery does not mean having a plan for the whole, but having an awareness of the whole.'[8] The pattern of reversal implies just this – that our primary task in enabling social organisms to emerge and thrive is not to plan, predict, structure and control but to facilitate the organism's awareness of itself, of its own identity and impulse, of what, in Pogacnik's words, is 'trying to be born'.[9]

Awareness, the ability to see, grows through understanding. This is the spirit in which this part has been written. There are other archetypes which manifest through social organisation, as there are patterns which have been created by us through the repeated relationships and events which we have embedded into social situations over millennia. The more we uncover and investigate such archetypes and patterns, the richer will be our appreciation of the beings with which we work. The patterns mentioned here are those that seem to me most relevant particularly with respect to change and development. But there is no end to what we can discover.

EXERCISE

Making Culture Conscious

A particular aspect of the principle of reversal, a way into understanding it from within, is the recognition that it is not only the world out there which configures our life as it is, but the configuration of our life which impacts as much on the world out there. As we unfold along the path of our becoming we develop a habitual way of doing and seeing things, a personal culture, which dictates much of what we experience in the world around us. Such personal culture is enmeshed in, and in many ways derived from, the collective culture(s) which surrounds us and of which we form a part. Yet it is also personal and unique.

A great stride forward in the move towards greater freedom, towards greater autonomy and authenticity, is the ongoing work we may do on making our own culture, our ingrained habits and norms, more conscious and transparent. In so doing, we learn what it is to work with the personal cultures of others and with the group cultures developed by social organisms. Thus do we begin to reverse the cycle of blame and complaint, and take responsibility for both our light and our shadow, for what we contribute and for the constraints we impose on others and on ourselves.

Exploring our personal culture, though, is not necessarily – at least initially – an easy or pleasant experience. If we are looking at the habitual ways of responding which we have built up over the years, and if, as the poet says, 'Failure is to form habits',[10] then uncovering our own culture will reveal its share of failures in our own development process. This is not unexpected; consciousness forms through pain overcome, and it is hard to imagine development towards greater consciousness which is not accompanied by feelings of loss and even depression as the process unfolds. It is in our acknowledging, letting go and moving beyond such patterns that the promise of development lies, not in avoiding them. These supposed failures are not failures at all but devices which once served as legitimate protection and aid, and which only over time may have become outlived and in need of transformation. Many of the habits we develop serve us very well, and will continue to do so into the future. But in either case, it helps to be aware.

This specific exercise runs like this. Consider two opportunities which you have taken in your life, and two opportunities which were presented but which you did not take. Consider too, two obstacles which you have overcome, and two obstacles which you were not able to overcome.

In each case, re-create in detail that which caused you to follow the course of action you chose. (Once again, a speaking partner will help to take you that much further, though solitary reflection before sharing is necessary.) Once you have outlined all these cases and the causes of your own choices which you are able to identify, the idea is to look for patterns running through the stories, and particularly in the apparent causes of your choices. What are the patterns which emerge, which link these stories into a living whole?

Once you have identified these patterns, ask yourself what this says about the things which you tend to draw towards you, and the things which live inside you. Where your answers to these questions begin to come together, you will be able to identify and describe aspects of who you are and of your own personal culture. Your understanding of yourself will deepen as you do so, as will your ability to empathise with the struggles which the groups you work with will experience as they work on their own change processes.

Part III

Change

The idea of the responsible interpretation of the world, of the individual equipping him or herself to become a conscious interpreter of the world, is central to democracy.

Gianni Vattimo

Observe how all things are continually being born of change ... Whatever is, is in some sense the seed of what is to emerge from it ... Make a habit of regularly observing the universal process of change; be assiduous in your attention to it, and school yourself thoroughly in this branch of study; there is nothing more elevating to the mind.

Marcus Aurelius

14
The Narrative Thread

[T]he single principle [is] responsible for every event or thing ... Even though the single principle cannot be defined, it is possible to explain what is happening in a group ... The wise leader returns once again to an awareness of the single principle that lies behind what is happening.

John Heider

Every living system is always in a state of change. Whenever and wherever we, as social practitioners, intervene into such a system, we are entering a story, a narrative, which has its own trajectory, its own meaning and integrity, its own truth. And that story, that truth, is constantly evolving.

We turn now to address the nature of change in a living system. Using a way of seeing which foregrounds relationships and the whole, and backed by a certain sensibility and appreciation for the patterns governing social systems, we must enter the actual process of change itself. It is at the coalface that acts of intervention are performed.

We saw earlier that not only is the whole enfolded in the parts, but also that these parts are themselves wholes – organs or organisms – in their own right. So, for example, the trees in a forest, the organs in a body, the individuals in an organisation. Each part, then, is a complex living system in its own right; each has its own meaning, each is a whole. Each is itself in a state of constant change, and it is this multilayered, multifaceted tapestry of simultaneous change processes which gives to living systems a quality of *dynamic complexity*.

The same input will get different responses at different times. Initial tiny shifts, seemingly insignificant, may, through positive feedback – because the system is alive and self-organising – generate new conditions which are wholly unexpected. On the other hand, huge inputs may be absorbed by the system through negative feedback, leaving it much as it was before. And these alternative responses are difficult to predict. So the mind, the weather, the economy, water, social organisms, life itself – all manifest a dynamic complexity which allows that to emerge which could never have been predicted from the component parts, or from a particular input to, or initial shift in, the system. Social change then, becomes dynamically complex, and not merely complicated.

In such systems, change does not occur smoothly or progressively, as in the linear curve of a graph; rather, change is often *discontinuous* – long periods of relative stability will suddenly be disrupted by a radical break with the prevailing immobility or trend of change up to that point. Stability is not a feature of dynamically complex systems; change is therefore unpredictable and uncertain. We cannot foresee accurately what the future consequences of present actions will be. It is often difficult to understand exactly what caused a particular change, or why a particular input realised very little change, or what the full impact of a particular action might be. Some systems appear 'ready' for change; others do not. Change is, in a very real sense, ungovernable.

Yet we know that there is also coherence – living systems do not simply become more and more chaotic until they fall to pieces. We know too that there are aspects that we can foresee, and prepare for, if not control – thus we are able to make sense, to an extent, of the weather, of the mind, of an individual or social organism. As the new sciences appear to be demonstrating, the very openness to change, the very chaos characteristic of complex systems seems to lead to new and higher forms of organisation. They appear purposefully loose, so as to be able to absorb outside influences almost as nutrients in their continual process of becoming. Such openness cannot permit of control from the outside, but this does not mean that it is out of control.

This relationship between chaos and order is a significant aspect of complex systems. Systems which are capable of growth and development exist in a state which has been described as being 'at the edge of chaos'. Where the system is loose and chaotic new elements are allowed to enter and form a new and higher level of order. Where there is rigidity this cannot happen and the system gradually atrophies, losing resilience and life. (We have recognised this when looking at the dynamic of power.) We can say that order emerges out of chaos, through the self-organising ability of complex systems, an ability which uses the openness contingent on chaos to form more complex levels of order.

So within chaos order exists as formative force (the kinds of patterns we have been referring to) which does not allow complete fragmentation of the system. Instead, such patterns act almost as benevolent boundaries which allow the system to reach the levels of chaos necessary for it to be open and adaptive; and through such adaptation to form new order.

The point about a self-organising system is that it seeks balance – while it can explode into far-from-stable non-linearity (utter chaos), it will always attempt to return. We know, for example, that the rich get richer, but if income differences become too great then the system may explode into chaos (riots, revolution). This means that the system has to adapt again. An entirely new form of social order – perhaps unpredictable and unexpected – must emerge. In the long run, the social cannot remain in a state of chaos, and patterns will assert themselves. If order were seen as one polarity and chaos the other, then threefold thinking may reveal the balancing (third) concept as 'life' – any move too far in the direction of the polarities will result in death, either by stagnation or by fragmentation.

Some of the patterns which provide order for social organisms have been described in the preceding section. However chaotic the system, however cataclysmic the circumstances, however unpredictable the future of the discrete parts may be, the whole will in the long run cohere into a recognisable form and process, guided by the 'non-changing' and the universal. Providing order, too, is the weight of history which lies on the organism, the identity which it has built, that which has been done to the point where patterns unique to the particular organisation or community have formed. Danger lurks, however, in the order which forms as the crust of past endeavour – while it may serve to provide meaning and identity, backbone and structure, it may also *encrust* the organism with stale habits and routines which allow little movement, and seal the organism off from the new which is borne in on the tides of chaos.

Chaos itself, on the other hand – unpredictability – is provided by the individuals in the system, those discrete parts which, while collaboratively forming the whole, are also wholes unto themselves in their own right. These individuals, with their own destinies, processes of change and becoming, and whose relationships form a matrix of potential and possibility, ensure that the organism is always heaving with change. They, as well as changing circumstances – the outside situation which shifts and turns – bring new information, new imperatives and demand new responses. Here too lies danger – too much openness will fragment and incapacitate the organism, leaving it prey to forces beyond its ambit and comprehension. While chaos may be admitted, and entertained, it helps to meet it with a certain tensile strength, which is dependent on the order prevailing in the system.

At any one time, the particularity of a social organism arises through the coming together of these two streams – order (underlying patterns and specific biography) and chaos (discrete individualities and the changing context). It is at the confluence of these two strands that the organism's development trajectory, its unique path, emerges as the whole with which we must work. Such whole is a flowing river of energy and power which influences all parts of the organism – which constrains and enables and provides significance for these parts – and, though it be invisible, it *can be read.*

It is this reading of the whole – which we have been referring to as a new way of seeing – which is requisite for a developmental social intervention. We read a life as we would a story. Our task is to uncover the narrative thread, born of chaos and order, and respond appropriately.

The dynamic complexity of the social situation makes it very difficult to work effectively with discrete parts, with isolated things – our efforts at effecting change fragment in our hands, have little or no effect on major variables, or too much effect on unpredicted and unforeseen aspects of the system. Working with discrete parts is to adopt a 'controlling' approach – we attempt to isolate and regulate pieces and sections because the whole is too complex; whereas we can hold onto a part, find its boundary, deal with it first before moving on to deal with the next thing. But as we do this we affect everything else in the system, and if we are not aware of the flow, awake to the underlying process, we can inflict great damage, or simply waste time. Attempts to over control, over intervene, or over structure, launched into dynamically complex systems, often cut across and dislocate longer-term self-organising patterns. The result is conflict, disorder or finally stasis. Of course we must intervene by addressing ourselves to discrete parts – there is no other possibility open to us; we cannot do it all instantaneously. But the point of signalling the danger of working with parts is to emphasise the need to ensure that any intervention we bring be congruent with the flow, with the need of the whole.

Reading the narrative thread, the particular dynamics of an organism's development trajectory or process, is demanding. Especially as so much of it lies beneath the surface, veiled and con-tinuously mutating, formed by confluences and intersections of overwhelming complexity. The intuitive faculty of the holistic mode of consciousness is needed, the ability to simultaneously perceive the invisible being which lies beyond the parts. Such consciousness is not primarily intellectual or theoretical or engineering or technical, but receptive and understanding. A reading of narrative thread remains supple, subtle and nuanced; it is iterative and gradual, building to intuitive grasp; it is reflective and reflexive (its accuracy must be tested and honed). Once again, we must trust our ability to see even as we develop that ability so that it may be trusted.

We penetrate softly, so that we can intuit underlying movements; we open ourselves so that the organism reveals its inner secrets to us, even as it reveals them to itself. There is a place where the organism wants to be revealed, for only in consciousness can it become fully whole and meaningful; through such intuition the organism becomes so much more than it was, even to itself, even as we ourselves become larger and more whole.

The narrative thread which we uncover is real – not a figment of our imagination, though a product of it. Ultimately, the ability to discern the narrative thread will be accomplished through the faculty of imagination. South African writer Olive Schreiner wrote: 'There is nothing so universally intelligible as truth. It has a thousand meanings, and suggests a thousand more.'[1] We are always approximating, and of course the story will be different depending on particular viewpoints and perspectives, but intuition and imagination can be honed such that they progressively enable the world (social organism) to reveal itself. And as it reveals itself, through our own participation, involvement and collaboration, our interventions are enabled to respond to the whole, rather than to our preconceived ideas and factory-built solutions.

The specificity of social interventions, the need for them to respond to the unique path and moment of the particular organism, is the mark of their effectiveness. Packages, preconceived solutions, programmes designed and delivered for an anonymous market, concepts and techniques presumed to be effective for all social organisms at all times simply because they might be effective with some – these interventions struggle to be developmental when they do not respond to an accurate reading of the life process of the organism; then, they can become fragmented attempts to control and manipulate discrete variables in the hope that this will solve the problem. There never was a problem. The art of social intervention requires patient observation and receptivity, so that the organism becomes revealed sufficiently to enable an accurate intervention to unstop the blockage, assist the flow, and guide the organism further along its own path.

EXERCISE

Discovering Wonder

The specific exercises which are contained under this heading are valuable in themselves, but this is about more than doing exercises. It is about inculcating a certain attitude of soul into our approach to the phenomena of our work. Attitudes cannot be trained, and they do not arise simply from the practising of exercises, though these may help. But attitudes can be understood and experienced, and drawn up out of the depths of soul; and once so drawn, they can be strengthened.[2]

We will never penetrate to the heart of an organisation's story without a deep sense of respect; the organism will not open up its innermost secrets to us. But even where the organisation does open up its secrets, the story itself, the narrative thread, will only begin to resound within our own beings if we approach it with a sense of wonder. With the feeling that there is a unique story here, a miraculous story. A story important for all the rest of the world occurring in its vicinity, a world which is influenced and formed through interacting with that story. Wonder enables us to see beyond the superficial, beyond the material; this attitude is as important as any other aid to observation. We can cut ourselves off from truth simply by denying the wonder which lives in what we see. Where we reduce or belittle that which is outside of ourselves, we reduce and belittle ourselves. Awe for that which is greater than us allows us to become greater than we are. Ennobling our world, and those we work with.

The following exercises may help. The idea is the building of peripheral vision.

Focus on an object in front of you – it can be as simple as a paper clip or stone, or as complex as a body of water or a painting. In the first instance, stare at the object with fixed gaze directed towards the details, and try to see detail after detail, separately; list them as you see them, and – because this is an exercise which should be performed daily over a number of days – each time you return to this focused gaze, try to discover new details. In the second instance, alternating with the first, while maintaining your focus, simultaneously expand your vision outwards, towards the periphery. Without relinquishing the centre, try to widen your gaze so that you are taking in more and more of what is happening to either side. We may call this peripheral vision – it is difficult to achieve and maintain, and raises entirely different feelings from focused vision. Gradually learn to do both, through alternating between them, and in doing so pay attention to the different feelings that arise in the soul.

After progressing somewhat with fixed objects, try it with moving objects; crowds of people, a flowing river, the flight of birds. Or try this: sit opposite someone – who consents to this exercise – and stare at him or her just above the eyes, in the centre of the forehead. Focus in on this spot. Once again, alternate this focused stare with peripheral vision: thus, without once relinquishing such focus, simultaneously expand your peripheral vision until it is able to take in the whole head and beyond. Concentrate on both periphery and centre at the same time. All these exercises may be repeated as often as you can (in this latter instance, with whomever will join you in the process – it's quite feasible to continue discussions and conversations whilst engaging with this exercise). Certainly a chosen phenomenon of observation should be repeated for a number of days. Slowly new observations will begin to emerge on that periphery, and even feelings and intimations which were not there previously.

This experience of the difference between focused and peripheral vision is a vital one. You can repeat such an exercise whilst walking in the countryside; as you walk, looking directly at the various things which will occupy your attention, keep expanding your peripheral vision as broadly as you can. Each time you do the exercise, see if you can expand it slightly further. Most important, though, is simply to become aware of the difference in experience between focused and peripheral vision. We can do the same thing with our hearing faculty (in fact the basis for such an exercise has already been outlined in 'Listening (1)', pp. 48–9). In these ways we begin to develop new faculties for observation.

It is on the periphery of our vision that the invisible can begin to be seen, for it is here that the invisible begins to mingle with the visible. It is on the periphery that the story of an organisation or a community will begin to make itself manifest. Staring at something too directly divests it of nuance and subtlety, as if colour has been drained by the harsh glare of the midday sun; only a flatness is left without that which emerges through the interplay of light and shadow (for shadow has been eradicated under the midday sun). Walk into an organisation and community with your eyes clearly focused on what is essential, but keep open to what you will see while looking out of the corner of your eye. It is here, in the realm of the unexpected, that the play of pattern and archetype and individuality will meet in the formation of narrative. Movement and process are discerned more accurately through peripheral vision; the underlying whole is more directly intuited. The difference between analytic and holistic consciousness is manifested in the shifting from focused to peripheral vision. Chaos enters from the periphery, and impulse; that which comes from without, from the future, and from the deeply inner underlying threads of destiny and possibility, enters from the periphery.

From out of the corner of one's eye, the shadow will first be perceived; the new coming towards us from out of the future will be glimpsed; the warp and weft of pattern may be sensed. Wonder enters here as well.

15
Stories from the Field

Creation is only the projection into form of that which already exists.

Shrimad Bhagavatam

Here follow a couple of stories, each comprising an attempt to read narrative thread.

We consider first the case of a small professional agency dedicated to the provision of developmental consultancy services mainly, though not entirely, to the non-profit sector. I had been called in to work with ... let us call it the Impulse Institute ... in order to help it develop a more competent and professional practice, given that the Institute was young, and that the consultancy staff were relatively young as well, particularly as regards their experience in the field. But there were other issues which needed to be worked with – conflict between consultants, undifferentiated and undeveloped work patterns, lack of clarity with respect to roles and function. I ensured that I had the time to listen closely to the organisation, to observe it closely, and to give the players themselves the time (and some of the necessary approach) to enable the organisation to listen to itself.

What emerged sounded something like this: the current director was not the original founder of the Institute. It had been founded many years before, by a consultant of such experience, depth and learning that he had attained an almost cult-like status amongst many younger consultants. Over the years many consultants had come and gone, trained and nurtured by him, and all of them, without exception, attributed their own expertise to his tutelage and mentoring. Some of these consultants had founded other agencies which were thriving. Yet the Impulse Institute had never really grown beyond that one founder – though people came and learned, they all left relatively quickly, and often through conflict. For all his genius, the original founder could not find it in him to work collaboratively and his institute limped along – really an institute in name only. And then, suddenly, he died without warning at a comparatively early age. At the time, the current director had joined the Impulse Institute and was under his tutelage, and he revered his tutor. Many of the other consultants now belonging to the Institute were also loosely connected with the original founder at that time and also revered him greatly.

Since his death – and by the time of his death, sufficient skills and expertise existed in the group – the Institute had been trying to build itself. This had been going on for some years, and certainly more consultants had joined their fortunes to the Institute. But in spite of attempting many different interventions – including vision building, strategic planning, conflict resolution, team building, restructuring with respect to roles and functions and decision-making procedures – the Institute remained somehow crippled, unable to thrive, and none of the consultants were able to really commit themselves to the agency beyond the next job.

Everyone associated with the institute knew these facts, particularly the current director, yet none had really heard the story before this particular intervention. No one had really observed the Institute, listened to its story as it had unfolded over time; no one had taken the trouble and time to really pay attention to the invisible patterns which gave rise to the phenomena manifesting on the surface. No one had thought of looking for the whole – all had been mesmerised by the parts, and by their attempts to remedy and control the dys-functionality of these parts. We tried this time, however, to penetrate through to the narrative thread which joined the parts into one whole. So the full story was told by various players, and everyone tried to listen in a new way. And the point of stuckness became accessible and transparent to everyone, even to the director, who till then had been a source of as much confusion as inspiration.

It became apparent that the original founder's untimely death had trapped the institute into a 'way of being' – there is simply no other way to put it – which constantly replayed his particular (and unconscious) strengths and weaknesses, without compensation for his lack of presence. His consultancy expertise had been inspiring, but he had never 'institutionalised' such knowledge into a replicable practice capable of being taught by others – thus he remained guru. He was open enough to allow anyone to come in and work alongside him, but always as apprentice – as soon as the apprentice turned into peer, conflict arose, the peer left and the Institute was unable to thrive.

Now the same patterns were manifesting – not least in the behaviour of the current director, not least in the fact that all consultants used the name of the founder ubiquitously when they spoke. As a social practitioner, my response was to point out that they had to 'let go' of the founder, which meant also diluting their devotion and reverence with a healthy dose of scepticism and honesty (for it was the devotion and reverence that was their undoing). Not only individually – although of course this as well – but the Institute, as social organism, had to put the founder behind it, and move on. Perhaps the name must change, certainly many practices must change, so that this newfound consciousness would not be lost.

And in fits and starts – as is the nature of transformation – the Impulse Institute did change. In the light of this new understanding, interventions which had previously made no impact slowly began to make a difference, and the Institute began to thrive, because it had become conscious of its own story, and was therefore free to begin to work on a new and consciously chosen future. The Institute had penetrated through to the shadow of its founder which was holding it captive, and was able to recognise the need to assert polarities which had previously been 'outlawed' (such as common standards, institutional accountability, procedures for developing a commonly accessible practice). Inspiration became grounded – through a process of letting go (literally allowing water to flow under the bridge) – such that, through such grounding, a new cycle of creativity could begin again in warmth and trust. The Impulse Institute was enabled to move further on its path of becoming that which the founder had originally intended, and which his death had at first hindered, and now – through awareness – was beginning to assist. Not without resistance and not without confusion. Thus chaos (the unexpected) met order (the forged pattern) to enable a story to unfold – and so may we make that story conscious, to make it ours once more.

As a contrast, let us look at an international development (or aid) organisation, with offices throughout the so-called developing world, head office in the north, responsible for vast sums of money, a major player in the countries in which it has offices. Working as a consultant to one of these country offices, staffed almost exclusively by locals, I am struck by the pervading organisational culture which spreads outwards from that head office in the north. In spite of the (current) absence of expatriates, this is very much a foreign, not indigenous organisation. It is suffused with the patterns of the parent, rather than with the dynamics of the local culture. Which is only the start of the story.

This is of course a very complex system, and we cannot do justice to it here. I raise only one consideration with respect to this country office. As I worked with it, as the months between the various contacts and interventions rolled by, I realised that, while much had changed for the better in the organisation, and there were new ways of working, new structures, new levels of effectiveness and impact perhaps; nevertheless a central dynamic – perhaps the most important dynamic of all – did not shift. The organisation was unable to develop a flexible and responsive approach to the problems and possibilities it faced. Somehow, it was always able to learn something through the interventions which we undertook together – it managed to take significant decisions about aspects of its organisational life, and change them. But once the change had occurred, it grounded the new way of working as rapidly as it could, in procedures and regulations which allowed for no innovation, no experimentation, no learning from or adjustment to this new way of working, no ongoing reflection and improvement. Once the decision and the change had been made, the new situation would remain as it was until crisis or extreme dysfunctionality forced the organism to ask its consultant to help it review once more.

So, when forced to, the organisation could learn, but it could not take on a learning approach. Qualities of air and water were foreign to it; the earth was its only medium. It could not begin to appreciate reversals indicated by the way of the spirit – which would allow openness and learning, to do less and accomplish more, to enable understanding to replace the technical manual, to encourage accountability to be exercised not through rigid bureaucratic require-ments, and not through instruments or an instrumental approach, but through the making of individual and collective *sense*, by individuals and by the various teams. Constant space to practise and innovate, and constant reflection and learning from that experience, are preconditions for effective development practice, but in this organisation they were entirely absent.

A biographical approach to the organisation made for a revealing exploration. Though it is now 'decentralised', with country offices relatively indigenised and autonomous, for a long time it was run by the central office in the north. The intention was certainly redress with respect to poverty in the south, but the manner of execution was patronising. Local staff were never given too much credit or respect, though they were always well looked-after and provided for. Local innovation and creativity and leadership were not expected; but proper performance of tasks and projects set centrally, and rigid procedures for accountability, were the order of the day. Over the years these things had changed, though largely in theory – in practice, in order to ensure accountability, the organisation had learned to set long-term plans, stick to those plans and report copiously on any deviations – regardless of the uncertainty surrounding both country context and organisational methodology. Results were tabulated in quantitative terms, and largely with regard to output. The quality of interventions was seldom reflected upon, and while questions of impact sometimes surfaced, little monitoring and evaluation was done with respect to actual outcomes. Reporting was not geared towards organisational learning and improvement but towards external and public auditing, so the organisation learned to produce vast quantities of paperwork – reports, tables, graphs, proposals, consultant commentaries, programme evaluations, requisitions and applications – which contributed almost nothing to its efficacy but which effectively tied people up in administrative knots for much of their working lives.

This was now changing, but the pervasive culture of 'approval-in-triplicate' was firmly entrenched. How then could the organisation begin to become more flexible, to adopt a learning approach to its work? Every move towards flexibility was strangled by a narrative which was largely unconscious, and all the more powerful for that. Staff drew pictures and developed metaphors and stories in their attempts to capture the narrative thread; all came down to a similar picture in the end – that of a herd of cattle, plodding one after the other behind the herder, and so habituated that they would plod that path were there no more grazing available at its end, and the herder had disappeared for the city.

Discerning the indelible pattern enables appropriate intervention. Local leadership had to be freed of this suffocating story, and in pursuit of this a programme of leadership development was instituted. Not a training programme, there had been more than enough of those; but a learning programme, or rather, one that focused on unlearning of patterns of behaviour which had become both ingrained and thoughtless. This programme has no set content, though its processes are rigorous: working in groups (case-study presentations and reflections, sharing of reflective reports, study of concepts and texts, in-depth conversations around issues and

futures, collaborative experimentations), working in pairs (developmental counselling, supervision, mutual accountability) and
working alone (self-development, journalling, ways of seeing, and
so on). In these ways individuals are gradually being freed enough to
be able to see the narrative thread from the outside, to tell the story
rather than only be told the story; to become as much narrator as
character. And gradually the organisation is freeing itself as the story
is retold, as the future unfolds to a different tune, and as individuals
begin to assert their own right to wholeness.

Although these stories concern organisations – and troubled organisations at that – they illustrate profound principles which manifest
themselves in all social organisms, including social initiatives,
communities, interest groups and social movements. The stories
which lie at the heart of social organisms, the narrative threads
woven from the warp and weft of order and chaos, pattern and
chance, are as manifold and varied as the infinitude of social
organisms themselves – but they are solidly united in the overall
story they tell.

As communities, groups and individuals we are both participant
in, and subject to, the vast array of social organisms whose activities
and presence saturates our lives – from multinational corporations
to governmental departments to the organs of civil society; from
geographic communities to ethnic groupings to political parties;
from gangs to churches to environmental conventions to chambers
of commerce. All of these manifest as autonomous beings of
immense power and influence, and when they operate from out of
a lack of consciousness of who they really are and what they really
are doing and effecting, then our entire society is at risk. And it is
only we who can bring them to consciousness – though we are also
implicate, and though we are also a consequence. Learning to
discern the invisible (but all too real) whole from the obvious parts
is not simply a consultancy skill; we are all social practitioners, and
this concerns us all.

That which manifests is only 'the projection into form of that
which already exists'. The invisible patterns give rise to what we see;
the story lives beyond and behind the book; the organism is a
vehicle for the spirit. Searching for the whole, we are searching for
meaning, and meaning cannot be quantified, weighed or measured,
just as the story cannot be reduced to an equation (however many
variables that equation may include), nor to a finite number of

discrete 'indicators'. Einstein pointedly noted: 'Not everything that counts, can be counted. And not everything that can be counted, counts.'[1]

It is the story, the inner life of the organism, which determines what it is, what it does, and how it behaves and relates. We are subject to these stories, these inner lives; it is these which govern our lives, yet for the most part we wander through life oblivious to the depth of the influences which pervade our communities. But how immeasurably richer life becomes when we open ourselves to seeing differently – and how much more powerfully we can participate in ensuring that the unfolding adventure holds hope for us all.

Of course not all narrative threads are seminal – there are a range of stories running through every organism – and not every unconscious narrative thread is negative. Consider the company which was facing ruin because it produced products which had become outmoded, yet did not know how to adapt because these products were all it had ever made, for generations – on searching deeper it realised that its 'narrative thread' had in fact nothing to do with specific types of goods or methods of fabrication but *general abilities* of the firm to *flexibly* carry out many different [activities] *at the same time*.[2] Armed with this understanding it was able to reinvent itself through realising its actual identity, rather than contriving to become something that it was not.

Or the rural community worker who, working with ruined communities whose lives had originally been held together by trusting relationships which had broken down through war and oppression, eschewed the many development projects which were now being imposed without regard for this underlying pattern of relationship which had been destroyed. He recognised that the projects were exacerbating the problem (lack of trust) through not recognising or respecting the underlying pattern which needed to be renewed if any future was to emerge. In his own words: 'The village is like a basket that has been broken and the pieces scattered ... What has been broken can be rewoven slowly and gradually, but only by those who will take the time to stay close to the village people and build trust with them ... Eventually the village people are the weavers themselves and they carry the task forward further, further.'[3]

When working in social situations we too often deal piecemeal, isolating different parts, looking for hasty solutions, needing to move on to the next item, regarding the situation before us as something to be surmounted. We're looking always for the how. And we seldom stop long enough to gradually become aware of the what – what is it exactly that lies at the heart of this conundrum which faces us. These narrative threads constitute the what – they describe the identity of that with which we are working, its inner being, the way it works and that which can be expected of it. To see the 'what' requires intuition; it requires that we first describe what we are working with (respectfully, else it will not reveal itself) before deciding how to deal with it. And it suggests that when we have understood *what* we are dealing with, we will know *how* to deal with it. As Julia Cameron, in her remarkable book *The Artist's Way*, puts it: 'Understand that the *what* must come before the *how*. First choose *what* you would do. The *how* usually falls into place of itself.'[4] A slightly different perspective, but the point is the same.

The point is borne out not simply in diagnosis but in preparing for the future. We know that over-planning, too much attention to detail, too much quantification and the fragmentation called for by logical frameworks and other planning instruments render organisations confused and alienated, fulfilling tasks whose point escapes them. Though perhaps it made sense when originally written, it fast loses meaning as the process unfolds. Whereas the organisation which prepares and maintains itself by making very clear what it is doing, and why, and how this relates to the inner core of both the organism and the issue being tackled, has little need of detailed planning with respect to the how, because the how falls into place of itself, once the inner pattern is understood. If we do not work out of the whole, then we work only with the pieces; and the relationships, which let us know where we are, are lost.

EXERCISE

Characterisation

The whole expresses itself in signs and symbols. This is indicated by the fact that it is 'no-thing', that it is invisible, intangible; perceptible only to a faculty capable of both attentive observation and imaginative insight, working together. When reading the whole, when searching for meaning, we make use of another language, a language which speaks in signs and symbols; the language of poetry, of myth and story.

Characterisation is a mode of observation which seeks to discover the essentials of a situation through describing it in its own language. The use of metaphor, incorporated into narrative, allows character to emerge; we begin to get a real sense of relationship and connection, of the invisible field which is being expressed, and of how things interweave to form new phenomena. The living character and significance of a phenomenon is revealed. As noted by the Jungian writer and therapist Robert Johnson:

> Our English word fantasy derives from the Greek work phantasia. The original meaning of this word is instructive: it meant 'a making visible'. It derives from a verb that means 'to make visible, to reveal'. The correlation is clear: the psychological function of our capacity for fantasy is to make visible the otherwise invisible dynamics ...[5]

Paul Matthews, in his book *Sing Me the Creation*, written to help develop the life of the imagination, uses the contrast with definition to approach a real appreciation of characterisation. The definition of a horse is supplied, as caricature, by Charles Dickens in his novel *Hard Times*: 'Quadraped. Graminivorous. Forty teeth, namely twenty-four grinders, four eye teeth and twelve incisive. Sheds coat in the Spring; in marshy countries sheds hoofs too. Hoofs hard, but requiring to be shod with iron. Age known by marks in mouth.' As a contrast, here is a characterisation from the Book of Job:

> Hast thou given the horse strength? Hast thou clothed his kneck with thunder? Canst thou make him afraid as the grasshopper? The glory of his nostrils is terrible. He paweth in the valley, and rejoiceth in his strength; he goeth on to meet the armed men. He mocketh at fear, and is not affrighted; neither turneth he back from the sword. The quivver rattleth against him, the glittering spear and the shield. He swalloweth the ground with fierceness and rage; ...[6]

This kind of approach is easily practised. We can take any phenomenon and attempt – through writing or speaking – to characterise it. You can try comparing the characters of two different kinds of tree, of the difference between suburbia and a squatter camp, of an intriguing relationship between two people. Think poetically, think as a narrator would, using metaphor and analogy, associations, representations, comparisons. Use myth and legend and proverb, the language of prophecy, the cadence of legend. Enter the world of the archetype, of quest and journey. Free yourself to write and to imagine yourself into the possibilities.

We can characterise, and help groups to characterise, their situation, a meeting, the state of their leadership, the relationship between men and women, or between administrators and field staff. Images and narratives will thrive together, build on each others' richness, evolve towards a central characterisation. We can characterise a group's shadow, or our own. Or its culture, or history, or striving. Through characterisation we gain a measure of the value of an intervention, the impact of a programme, the development of a relationship. These things are not as available to any other means, because no other means so speaks the language of the whole.

We can use characterisation in our work with each other – discovering each others' questions, aspirations, understanding of each other. So do we get to know each other, come closer to our own meanings, realise our collective responsibilities. And help groups to do the same. Characterisation enables the invisible patterns and movements of a situation to reveal themselves.

When characterising someone's story, or a situation, try to respond first by acknowledging where that story has touched you, where you have resonated with it. Such openness enables the real work of characterisation to begin, because it recognises that there are aspects here which may be mainly in the listener rather than in the story. By making this conscious we are freed, to an extent, from the projections of shadow.

Characterisation exercises develop the capacity to build longer stories. Moving from one image to the unfolding of a narrative, further depths within the whole may be discovered. The process of an intervention, the transformation of an aspect of community life, the impact of an action, the relationship between various forces played out within a particular context – all are accessible to narrative. Not least because narrative, unlike the dry factual report, embraces (rather than seeks to diminish) the presence of ambiguity, contradiction, and the reversal of conventional logic. And in so doing, enlarges possibility.

16
Metamorphosis

One does not discover new lands without consenting to lose sight of the shore for a very long time.

Andre Gide

Each of the composite beings mentioned in the previous chapter have many stories running through them. Like intersecting ripples caused by pebbles dropped into water at different points, they spread out from different centres, reinforce or diminish each other, create competing turbulences, affect the pattern of the whole. But beneath the waters, the tide pulls and tugs between moon and earth, and waves build and retreat in response. In every organism there are lesser stories and greater stories – and then there is the confluence of the whole itself, expressed as constantly evolving narrative thread.

This interplay between dynamic complexity and unitary simplicity is apparent when looking at social organisms of some stature, size, history and complexity, but it is equally true of small organisms. The board of an organisation that meets together three or four times a year; a study group that meets weekly or biweekly; a sports committee that meets every now and then; an environmental advocacy group that is formed of individuals from a number of different organisations; a networking committee that is made up of representatives of collaborating groups. Even meeting only three or four times a year, a narrative thread arises through the course of such work, and a living being is gradually born. Such groups assume particular characteristics indicative of their process of becoming.

We know, when we enter groups as regular participants, that they differ; our experience of them differs. Some we look forward to, knowing we will draw energy from their clarity, or enthusiasm. Others are all uphill battle, and a kind of nausea grips us as we contemplate the next gathering yet again. Though we and others are participants, there is also the character of the composite social being itself, which we enter each time. Accessible to us if we but pay attention. We can exercise leadership, work as social practitioner, wherever we find ourselves; every intervention may enhance the social fabric. In the words of St Francis of Assisi: 'Rather seek to understand than to be understood.'[1] Such an approach will carry us a long way towards helping the social organisms of which we are a part, to make sense of themselves, to penetrate their contradictions, to assume responsibility for their effects.

When it comes to nascent social organisms, if we are awake to these things, we can become aware of the patterns as they begin to form the strands which then cohere into the narrative thread. We cannot control them or predict what they will become, but we can take responsibility for what we bring to them. If we start with the knowledge that an entire story is just beginning, one which will affect all other stories, then the richness for which we are responsible becomes palpable.

Many of the wholes with which we work as social practitioners have already become distorted and unharmonious, biased or blocked, antagonistic or irresponsible, by the time we get to them. When we enter from the outside, we are called in precisely because of such stuckness, so this is to be expected. The dynamics we meet will often be quite debilitating. This is not always true. But it helps to acknowledge that many social beings demand concerted effort of anyone who would help them develop.

The organisational culture (or story, or whole) which has progressed for too long without helpful intervention, which has sunk beneath the surface into unconscious habit, will frequently display constraint and debility. Paralysis or stagnation, denial and avoidance, projection and blame – all consequences of 'unconscious drift'[2] – find their expression in the organism's (and the individuals') resistance to change. Once patterns have settled into a familiar groove, even where that familiarity is fraught with pain and discomfort, the organism will strive to maintain the groove.

The familiar is more comfortable than the unfamiliar. We know what to expect from the familiar; we know nothing of what to expect from a way of being which, however alluring it may sound, we have not yet experienced. Our experience is more powerful than our imagination. There is a dependency syndrome at work. Like a drug addict, we may agree that we need to change, yet succumb to the first temptation. Scared to let go without having something new to cling to, we cannot access the new until we let go – so we cannot let go. We are addicted to the clinging itself.

We could say that when we are stuck in a groove, we have become trapped by fields which pattern our very way of being; very little that is new is allowed to enter. The balance between chaos and order has been lost. The organism forgoes consciousness in order to protect itself – its very rescuers are seen as enemies.

D.H. Lawrence gets to the heart of the issue in his poem 'Song of a Man who has Come Through':

> What is the knocking?
> What is the knocking at the door in the night?
> It is somebody wants to do us harm.
>
> No, no, it is the three strange angels.
> Admit them, admit them.[3]

What happened, between the lines of the two stanzas above, to enable the man to 'come through', to finally understand that those knocking are not wanting to do harm but are there to provide possibility? That their very otherness is indication of help rather than hurt? How did the switch occur? This question burns constantly in the heart of every social practitioner – and perhaps is never finally answered. Perhaps the answer is not the key; perhaps the key lies in possession of the burning question itself. Lawrence does not attempt the answer; there is a movement between two entirely different worlds of relationship which he expresses simply by a line omitted, by the honesty of a resounding silence.

The moment of metamorphosis remains enigmatic. If we are to work with social change and development, with transformation and transition, then it may help to form an approach to that place of silent mystery. Where butterfly wings are grown within the shroud of the caterpillar's concealment.

We cannot create the movement which occurs there. Perhaps all that we can hope to do is to hold and protect, provide the conditions for the possibility of such transformation; boil the water and clean the house, so to speak. The birthing itself must be done from within. But we can help to make aware that the time may have come.

We know that the whole can be read, and we know that it resists being known. The whole prevents access to that which is moving in its depths because it fears that once the thread is uncovered, the only alternative will be change itself – always daunting. On the other hand, the whole is in a process of becoming, and there is deep (though often unconscious) acknowledgement that the more that is known, the more that process of becoming will be enabled to unfold. Through the struggle between these two tendencies the crises of change arise – those upheavals and seismic rumblings which shake the foundations of social organisms, often to their very core. That struggle between resistance and surrender.

If we are to enable the organism to reveal itself to us (and ultimately to itself), in the manner of Goethe's 'open secret', then we should not storm the gates. Too intrusive an approach will inevitably raise the defences – and why not, for why should the organism trust, when it cannot trust itself? Trustworthiness must be proven first, by the practitioner; must be offered and given first, not as part of a bargain but freely, without reciprocal demand. Then silence and patience and the sacred arts of listening and observation.

If we open ourselves to receive then we will be offered, because the world wants to be known. It does not help to wring the knowledge out, extract and cajole and attempt to achieve our own greatness into the bargain. Situations reveal themselves when they are met with real receptivity. As is it finely expressed by Bernard Lievegoed: 'Our task is not to judge, but to help and heal. We approach each situation with awe and wonder.'[4]

For those processes saturated with the very human attribute of self-consciousness, the process of metamorphosis – unpredictable and ungovernable as it is – has at least been partially described by the work of Elizabeth Kubler-Ross on death and dying.[5] Her work with terminally ill patients and their families has provided an illuminating account of the way we deal with loss, any loss. And because loss is so much part of the process of change, because any growth and enlargement is accompanied by loss and death, her work provides an important angle and approach with respect to the dynamics of transition.

Briefly put, our response to the crisis of impending death and loss runs through a number of stages. On first discovering, or being made aware, the response is one of denial – that this fate might belong to another, but definitely not to me. When denial is not able to change reality, then we tend to bargain (with whom? ... with God?) by trying to make some kind of deal: I realise that this is happening, but if I change in some relatively painless way might this not be enough to alter reality? This does not work either, and then anger sets in – why me, why not someone else, what is it about me that causes the world to behave in this way? When this does not work, a state of depression often sets in, a sense of being trapped – there is no future, and nothing can be done about it. With depression, a loss of energy.

Paradoxically, at this point – though our exploration of the pattern of reversal should prepare us – such depression may not lead further down but may instead be the prelude to moving beyond, to resurrection. For the step beyond depression is acceptance, acceptance of one's fate, one's loss; and here too is paradox. Acceptance would on the surface of things imply acquiescence, and continuing depression. But in fact, acceptance appears to lead to a kind of rebirth of energy, enthusiasm and particularly meaning, while the reality of death and loss does not diminish. Instead of being weighed down by loss, such loss takes on meaning and heralds new ways of being.

Always respecting that such sequences never occur with the precision and linearity of presentations such as these, this is the path that the social practitioner will tread on the journey which accompanies any social organism through metamorphosis. So with respect to responsive intervention, reading the narrative thread is not the whole story. *How* it is read is as important. How such reading is presented and worked with is equally so.

We are facilitating passages through denial, bargaining, anger; all of which will reverberate throughout the system in different ways and different places at different times. How to lift the organism – and those parts with which one will work day to day – beyond its depression? How to challenge and confront, to galvanise energy without forcing a retreat? How to discriminate between real and spurious forms of acceptance? How to capitalise on the energy that arises through genuine acceptance – and how to help those parts which have achieved acceptance without losing heart because other parts still resist?

EXERCISE

Turning Points

One of the most effective ways of understanding the process of metamorphosis is by experiencing it within our own life. The exercise which follows is a self-reflection exercise on our own periods and processes of metamorphosis. As with the other exercises of this kind it will be helpful to do this with a speaking partner, someone trustworthy and in whom you have confidence, for the exercise will take you into intimate and vulnerable aspects of your life.

First, identify three major turning points in your life, which may have taken place at any time in the past (nothing that you are currently in process with now). These are events or periods when your whole world was turned upside down, and when something so significant changed in your life that it shifted into another gear. A turning point without which your life would have been far different.

Describe each story in detail, such that your speaking partner can experience it as if he or she were there; a vivid and colourful narrative. Describe the circumstances and the people involved (if any); what precipitated it, how did circumstances and people's involvement change as time went on, how long did it take till you were through? Describe the events as the process unfolded. Describe your own responses, and how they shifted and changed, your resistances, your acceptances, your doubts and questions and certainties. And what changed for you, in what ways did your life change, as a result of this turning point?

Once you have described three stories in this fashion, look – together with your speaking partner – across all three stories, and search for the patterns of change and metamorphosis which you may discover manifesting across these stories. There will be patterns which will be unique to you – the way you personally approach periods of transformation – and there will be patterns which you share with the rest of humanity, patterns of a significantly archetypal nature. (So it helps if your speaking partner also does the exercise, and even more if there is a group of such pairs doing it within which learnings and insights – though not the stories themselves – can be shared and compared.) In this way we enter directly into the archetypal patterns of change and metamorphosis which affect all (self-)conscious beings, and so develop further our ability to intuit and work with such processes as they occur for composite social beings.

A further development of this exercise, which can be used to help practitioners to a greater understanding of the process of intervention itself, is to ask pairs to consider, once they have articulated the patterns from their stories, the following questions. When you were going through these periods of metamorphosis, what kind of approach would you have appreciated from someone who was there to help you; how could they best have intervened to facilitate the unfolding of your process? And, considering the stories of your speaking partner, what do you think you might have been able to do to ease your partner's passage through these periods had you been available at the time to help? The answers to these two questions will begin to build considerations for the practice of social development out of your personal experience of change.

17
Guidance

Our life passes in transformation. And the external wanes ever smaller.

Rainer Maria Rilke

As social practitioners, we deal with the evolution of consciousness. That which evolves has no predetermined end point; what emerges is unpredictable, irreducible, organic and new. The end result of a developmental social intervention cannot be decided beforehand. Working in open space as we do, to enlarge possibilities rather than narrow potential down to a planned result, places severe demands on the practitioner. We are called upon to practise a discipline within volatile, open space which we cannot control and upon which we cannot impose.

The only way to do this is to work from the inside out, as it were. If we cannot control from the outside, then we must learn to work in a way which corresponds with the inner (invisible) movements of the organism. Not to impose upon the organism our conception as to what must be, but to enable and guide that which is already moving and struggling to be born, to manifest. To respect the inner life of the organism, and to practise a discipline which corresponds with the patterns and forms which guide this inner life.

Bernard Lievegoed, referring to Rudolph Steiner's *The Philosophy of Freedom*, talks of a particular quality which the social practitioner must cultivate.[1] In order to be involved in, or to be a guide to, the transforming of the present into a future state – the essence of our work – one must have the skill to do this in such a way that the object with which one is working is not violated, but is *transformed according to its own laws*. 'In social terms this means that whenever one wishes to act in the social realm, one must always start from the premise that one's actions should not prejudice the freedom of others.'[2] Lievegoed calls this *moral technique*. And together with moral technique goes moral imagination. Moral imagination, as he describes it, is the ability to think new thoughts about the future, not arbitrarily, but out of an understanding and grasp of the principles at work in the present. In other words, to understand (see) *what* is at work, so that transformation may occur naturally, according to the already underlying process working through the system. Moral imagination is the ability to see the *what*, while moral technique is the ability to guide the *how* in accordance with the what. Thus Lievegoed notes that moral technique entails the ability to seek continually new social forms for social organisms, appropriate to each situation, in such a way as not to encroach upon the freedom of any member. Thus to enhance individual freedom whilst ensuring that the integrity of the whole, on its path of becoming, remains intact – as Lievegoed notes, 'an extraordinarily difficult task'.[3]

As we learn to 'read' development in this way, we realise that the skills and competencies which we require are not of the kind which can be taught on training courses. We are having to deal with ambiguity and paradox, uncertainty in the turbulence of change, new and unique situations coming towards us from out of a future which we have had little experience of as yet. The implication is that we need to develop a resourcefulness out of which we can respond, rather than being trained in past solutions, in fixed mindsets and behaviours and techniques which replicate particular patterns and understandings instead of freeing us to respond uniquely to unique situations. Moral technique cannot be trained; if we are striving for the freedom of others then we must work out of such freedom ourselves. But while it cannot be trained, it can certainly be learned – which requires continuous self-development, self-reflection, as well as mentoring and peer review possibilities. For social practice takes us into a realm of invisible qualities of soul, rather than a realm of physical quantity; thus we must cultivate, more than anything else, a sensibility for social phenomena, an intimacy with social process.

We should be able to hold the tension generated by ambiguity and uncertainty, rather than seek immediate solutions. Too often organisms 'jump to conclusions', they seek desperately for hasty ways out of current dilemmas, often before they really understand the 'what' of their situation, and certainly before the 'how' is enabled to emerge from the what. People, and often the organisms themselves, find it very difficult – indeed, painful – to sit for long periods with ambiguity, to live without resolution. We want to eradicate the pain and move on. But development takes its own time; while we can facilitate and guide a movement towards resolution, that movement should be able to take its own time – for we are dealing with living beings in a process of unfolding.

The process may be painful – periods of confusion and conflict, darkness and gradual gestation, experiment and failure. The temptation will always be to try to move faster than the situation allows – and the temptation will in many ways be justified, for why else are we there in the first place, if we cannot assist? Yet we are there as guide only, we cannot impose or create, so however much we wish to move the situation, the cue for movement must always emerge from the situation itself. If we move too quickly we risk imposing a false solution. The ability to live with such tension, within ourselves, is one of the more challenging aspects of social practice. We learn to work deeply, but with a light touch.

We must be capable of both hard and soft interventions. Soft interventions are primarily supportive, they draw the organism out, encourage it to emerge into a new future. Hard interventions are challenges and confrontations, which can be used when supportive interventions do not help in moving the organism due to resistances which have become too entrenched, and habits which have become too unconscious.

Process consultancy, or social development practice, which involves intervening into organisational systems in order to help right whatever wrongs may be manifesting, is often compared to a doctor–patient relationship, in which the doctor diagnoses the illness and attempts to restore the system to health. But for various reasons the analogy does not hold. For the doctor–patient relationship is often disempowering for the patient. The doctor diagnoses and treats, often without the patient being fully aware of the issues, nor involved in their own cure. Obviously this does not describe the point of social practice at all, which depends for its success specifically on a changed relationship with self and context, such that the organism has more authority, not less. But the analogy founders for deeper reasons as well.

In the first place, *there is no final diagnosis*. There cannot be, because the more you probe, the deeper you go, the more will be revealed. It is possible that the depth of the whole may be infinite; it may be that one can never finally plumb the narrative thread entirely. Because as you delve deeper, so more is revealed, as much to the organism as to you as practitioner, and as more is revealed the narrative thread changes, because the whole is evolving as you probe – indeed, through your very probing. The situation is changing all the time. Perhaps there is no final narrative thread, no ultimate diagnosis which may be made and adhered to through the resultant change process. As the process of intervention evolves, so a new narrative will emerge. And the process of intervention includes questioning, and the forming of relationships which will affect the whole. So the intervention itself becomes part of the story, part of the whole, and in this sense any diagnosis is always only partial, and ever emerging as the organism unfolds through increasing awareness of itself.

In the second place, *it is not only the intervention which becomes part of the story, but the practitioner as well.* There is no spurious objectivity here. The organism is already changing in response to a new relationship, which is the relationship between the practitioner and itself. Practitioners 'interfere' with precisely what they are researching. They become part of the system through the new set of relationships which is generated.

And yet, our assistance depends for its efficacy on our not being part of the system, so that we may work from the outside, see beyond what can be seen from the inside, bring some worthwhile added perspective to what is already known, operate from outside of the unconscious patterns which might have taken hold. But we are simultaneously inside, part of the whole which we are trying to help develop (and which does indeed exist independently of us, and which will continue its life journey without us). This contradiction with respect to our role – with respect to the very foundation of our craft – is not one which can be wished away. We are at once inside and outside, at once part of and not part of; we can help tell the story but we are also part of the story (which may be the part we leave out). One consultant described it as like being on the rim of a cup, neither in nor out, yet both in and out.[4]

Not so difficult to appreciate, really, for we have already explored the laws of spirit which entail not either/or but both/and; contradiction is the very life blood of the development process. But easy though it may be to appreciate, it is correspondingly difficult to achieve, to live with. Nothing is ever quite what it seems, yet we must work with what we can see, as if it is the whole story, always knowing that every moment the story is changing, that there is always more than we can see, and that often the mote in our eye which prevents us from seeing is our own presence in the intervention – and this we cannot ever be rid of. So our professionality must lie in this: that we do not deny, but embrace the contradictions inherent in our craft, and turn these into our very strengths. We are not doctors, neither are we therapists – though the analogy is perhaps closer here, for the therapist never presumes to have plumbed the depths of a client's psyche. We are something entirely unique, and we therefore depend for our effectiveness on the depth of self-knowledge which we bring to bear on the stories with which we work.

We are both participant and observer, both neutral and involved, at the same time intimate and distant. And we are always part of – and must take cognisance of – three distinct processes. The process of the organism with which we are working; our own process; and the process of the organism or culture out of which we come. None of these processes can be ignored, and they all affect each other. All are wholes – struggling to become – in their own right, and together they form the whole which is the intervention. The responsive intervention, then, is also in a process of dynamic change, as wild as that into which it intervenes.

EXERCISE

Listening (2)

We have considered the development of the faculty of hearing itself; we now concern ourselves with developing the ability to listen to a social situation. To listen for the essence of the situation, for what is really trying to be expressed. There are many listening exercises; here we simply table two, followed by one further brief consideration. All these exercises must be performed in groups, either small groups on their own, or as subsets of a larger group which will process insights and discoveries in plenary.

In groups of four, one person presents a problem which they are experiencing. Five or ten minutes of talking will do; and there is no interruption, the other three are only listening. The person presenting the problem must present an issue which is very real for them, and which is current and still unresolved. (We cannot fabricate a situation if we are to be true to this exercise; the problem presented must be real, current and unresolved.) The three people who are listening must each engage on a different level, and only on that level – the exercise may be done four times and participants rotated each time so that each person experiences every role.

The first of the three listens only on the level of content, on the level of thought – what have you heard that has actually been said, is your hearing accurate and your memory good, and can you summarise at the end of the presentation of the problem in such a way that the person presenting the problem feels heard? The second of the three listens on the level of feeling, of emotion, on the level of the heart – can you hear what this person is really experiencing as they consider the problem, can you empathise without imposing your own response? Can you give that back to the presenter in a way which allows the person to feel that you have entered into their situation from the inside, appreciatively, sensitively and with enough experience of your own that you can appreciate the emotions which you have listened for?

The third person listens on the most difficult level of all, on the level of the will, the underlying intention. There is that within the person who is dealing with the problem which already knows which way it is going to act; the level of the will, of underlying intent, lies in the realm of the unconscious (in the stream of life, as it were) and so the presenter is perhaps not aware of that which you – as the third listener – are listening for: the essence, the underlying whole, the character of the situation which will point towards the path to be taken into the future. Can you present this insight back to the presenter in a facilitative way, remembering that the person is not necessarily aware of this intent themself, that you are dealing with deep underlying levels of meaning and intent, and that you may well be wrong, because this level of listening is difficult in the extreme, and needs extensive practice. You are listening here for the invisible, the intangible, that which is struggling out of the unconscious to become conscious, to be heard in spite of itself, to express itself to someone who is prepared to listen (with wonder and respect) for the whole out of which the individual acts.

All three levels of listening are important, whether listening to individuals or social groupings; normally one will listen on all three levels at once, but this is an exercise, and we are separating them out in order to understand the levels individually and to practise each one separately so that we may differentiate between them, searching always for the integrity of the whole picture.

In groups of three, allow one person to present a problem – real, current, unresolved. The second person listens and also intervenes, trying to get clarity, asking the kinds of questions which will take the presenter further, trying to help the presenter to hear what is really happening on the level of underlying will and intent. In other words, a kind of counselling engagement, a facilitative process, with the emphasis on listening and asking questions. The third person listens to the second, and when it is over facilitates a process in which feedback is given to the listener/counsellor, particularly on their listening skills and on what may have helped and hindered and on what may be improved, and whether the listener's interventions – through questions, perhaps – helped or hindered their own ability to really hear the situation. (Once again, this exercise can be done three times so that each person gets to experience each role.)

One last point. We try to efface ourselves in these listening situations, so that we and our possible projections do not get in the way, so that we can hear clearly. But we do also have responses, and these responses are not always simply our own noise and issues; they may be genuine ('objective') responses to what has been heard, and then these responses should be used as authentic and legitimate aids to further understanding what is being heard. So, for example, if you are listening to a story and you feel anger rising within you, even though no anger is being expressed by the speaker, this feeling of anger may well be a key to hearing something real that is taking place beneath the surface, which the speaker has not yet gotten in touch with. Your own inner responses as listener may be the key to hearing on the level of the will. Or they may simply be projection, and cloud the real issue. The more we grow as practitioner, the more we grow in our understanding of ourselves, the more we will be able to discriminate between these possibilities.

18
Rites of Passage

There was something inherent in the necessities of successful action which carried with it the moral degradation of the idea.

Joseph Conrad

We intervene into the process of creation. And creation entails the gradual translation of idea into action, the incarnation of thought into material manifestation. It is not possible to carry an idea through to successful implementation, to incarnate a thought, without compromising that thought, without distorting the idea. Implementation, while it carries with it the realisation of the idea, carries also the intimation and foreshadowing of death; the creative moves from the fluid outpouring of impulse to the sclerotic residue of execution. There is no escape from the process of incarnation – pure spirit enters the material, and slowly the expedient needs of matter take over from the naivety of original impulse.

When ideas are not realised, they remain inspirations, leading towards the future; when they are realised, they eventually become the past, fraught with the mistakes and misguided attempts at maintaining and defending perfection. Successful implementation demands that we bow to the needs of expediency – there is no other way. The burgeoning of the seedling becomes the weight of dead wood; yet without incarnation the magnificence of the tree would be forever denied. And even as the life disappears, a seed remains behind, and in that seed idea and life are held anew, and the process of evolution continues. *When we engage in the practice of social development, we intervene into the deepest recesses of life.* We assist the unfolding of the creative impulse.

Every social organism exhibits the same creative round – as described in Chapter 9 – from seed to wood and back to seed, from idea to compromised implementation leading again to the drive for renewal. The unbounded idea, the inspiring vision, is finally replaced by the policy, or the manual, or the law; and a renewal must take place to thaw that which has become stuck, so that we can move once more.

147

We have noted before that this 'creative cycle' is the archetypal pattern by means of which an organism is enabled to realise itself, a fourfoldness which provides also direction and intent. So that we may facilitate the process of change from within, rather than impose supposed solutions from without, the discipline of facilitating such change, the process of intervention itself, must reflect the movement contained in the creative cycle.

Not that we can approach such processes in a linear or sequential fashion. When intervening, we may enter and leave the social organism at various points in the cycle, depending on the nature of the work and the nature of the situation. One may complete only part of the whole cycle, yet succeed in an effective development intervention. One may find oneself engaged in different steps of the process concurrently. One may engage in them in a different sequence, or in a number of different sequences at the same time. At the very least, we have to recognise that all these 'phases' will only ever be partially completed, because a development intervention opens things up, rather than closes them down; and because the social organism is in a constant state of becoming. Every intervention into any phase in the cycle will only ever be partial. And often we only know what we are doing when we get there.

The reality of a social practice lies in its very turbulence. But this does not mean that a disciplined approach to the facilitation of change cannot be useful – particularly when that approach is built upon the archetypal pattern underlying change. Such an holistic approach enables us to situate our interventions with due regard for the integrity of the social organism's ongoing process. The danger of social practice as a responsive discipline is that it can result in the indiscriminate use of a wide range of fragmented interventions, often one-off interventions unrelated to each other. An approach grounded on the deeper appreciation of the underlying 'non-changing' patterns of change can guide us to more authentic and congruent interventions.

There is a fine line between the use of frameworks in service of an organism's interests and imposing models as the only contribution which a practitioner has to make. No model can tell us what exactly is happening at any point in time and therefore no model can tell us what to do. I suggest here an approach only. But the approach can act as a map with which to plot the next step of the journey or to help find the way when the process appears lost, stuck or without direction. The approach can help a practitioner to know where they are in the process; we must understand where we are fitting in, where our intervention is located, where it can go from there. Nevertheless, this approach is nothing more than a map. It is not a blueprint, a technique or a policy. Nor is it the journey itself. This is formed together with the social organism in the moment of the intervention itself.

Having said this, let us look at the elements in the creative round in slightly more depth, as they relate to phases of intervention. It does not help, though, to be told what to do; we must develop for ourselves a deep sense for the elements and what the approach as a whole may mean. And then work creatively, out of that sense, to form unique responses to the situations which are coming towards us.

Warmth, the element of fire, has to do with relationship. An atmosphere of trust and mutual respect enables a true process of transformation to begin. This is not the case only when the practitioner enters from without, in response to particular need. It is also true for members of the organism – whether you are leader or participant, your ability to work constructively with the whole will be in direct proportion to the levels of trust, honesty and transparency which are formed and maintained between yourself, the other individuals and the group as whole.

This work involves the very human dimensions of personal warmth and integrity, self-knowledge and clarity. Clarity with respect to the forming of mutual agreement around role and function, method and process, objective and purpose, criteria for later evaluation and assessment, duration of intervention and financial considerations. Use of these different categories will vary depending on the formality or informality of the intervention, and will also depend on how separate the intervention is from the rest of one's relationship with the organism.

Developmental social interventions open things up. They encourage the process of becoming, of going deeper, of unlearning, of expansiveness. They release energies which have been constricted, allow things to be seen and said which have never been permitted before; and the development process respects only its own sense of innate timing, it cannot be planned to order. Yet we do need order within the process of intervention – it is precisely the boundaries which the practitioner creates which enable the safe space within which people, and the organism, feel free to let go, to take on the new. Some form of mutual agreement around boundaries is essential if trust is to be maintained throughout the volatility of the intervention. At the very least, what is required is clarity about role (what can be expected of the practitioner) and some form of understanding about purpose and objectives.

So to the element of air, the drive to understanding. Once relationships are established (but of course these things are not really sequential, and gaining understanding must begin from the moment of first contact). Once, though, relationships are sufficiently formed, then real observation begins. Many methods are available, different mediums of observation, but to reach real understanding about a social situation we will work largely through questions. Partly this means simply the cultivation of a questioning attitude – one of real interest, a quizzical respect, a genuine wonder at this organism which may be similar to others but which is in fact entirely unique. This is the attitude embraced by the Sufi poet Rumi when he writes: 'Sell your cleverness and buy bewilderment.'[1]

But more, we must be capable of actually finding those questions which can help us penetrate the barriers which the organism will set up in its resistance to change. Asking questions is thus a competency to be cultivated – the ability to ask an open-ended question which is not simply a shot in the dark but which is focused and challenging enough to facilitate honesty and new insight on the part of the organism and its members. There is little to rely on outside the cultivation of the art itself.

Part of the job is to unleash the real questions sitting at the heart of the organism's current situation, which are often far deeper than those with which we arrive. The idea is to take the organism deeper, behind its own defences, so that light begins to emerge from darkness. And it is not merely for us as practitioners that light must emerge – we are nothing more than sense organ – but for the organism itself. As sense organ, our task is to enable the organism itself to listen for, and understand its own story, and come to grips with the questions which will act as spur to its development process.

Thus to fluidity, the element of water. How can we help to dissolve rigidity and bring the system into movement? We identify two interventions here. These may occur as one unbroken whole – for example, in a single meeting or in a workshop – or through many contacts over a long period of time. (In larger systems there may be certain parts which move and change at a different pace to others; one will always be working differentially throughout the system as the process unfolds.) The first intervention is help with *achieving acceptance*. Somehow the organism's story, and the essential questions which emerge from that story, must be owned. The task is to facilitate people and the organism through resistance to the point where acceptance is reached, the point at which the organism is able to face its own story, its own shadow perhaps, even its own light, its own doubts, certainly, and its own questions. Face these honestly, and receive that paradoxical energy which flows with the onset of acceptance.

The second intervention helps to *resolve the future*. Here the practitioner must assist the organism through resultant change processes. This is perhaps the most difficult point of any intervention – the change process itself, the conscious choosing of a new vision, perhaps new goals, sometimes radically revised strategies, new ways of working and patterns of relating, new structures and policies. In the midst of chaos, the organism has to make decisions about how to approach an unpredictable future. New practices must be found, and a process of change prepared, which will take the organism through the phase of transition to a new level of functioning. The larger and more complex the organism, the more staggered and stuttering the process of transition will be.

Necessarily, a hardening and structuring needs to take place, to enable the organism to emerge from its phase of transition, to enable the new order to settle. So to the task of grounding, the element of earth. Meaning must be translated into form and structure and standard. The fluidity of the transition must be grounded into continuing organisational practices.

This 'final' phase may take the practitioner into many different facets of the organism's life. As we have seen, some time after the new has become enshrined within new standards and practices and procedures, it will lose its cutting edge. The time will come for new movement to take place. That time may only emerge long after the practitioner has left. But as we work on grounding, some things may be borne in mind. It helps to assist the organism to guard against the onset of such fixity too early. There are organisms which establish forms of grounding in which the fluidity of the creative cycle becomes a way of being, such that ongoing consciousness is ensured. In all such organisms, something of the following occurs.

Instead of structuring the organism with the kind of formality which attempts to predetermine the way people interact with each other – rigid hierarchies and set procedures – the organism is designed around a *situation-driven* conception. That is, the task at hand assumes the most importance, and the organism and the people in it are freed and supported and encouraged to work together in the best possible way – for each particular situation – such that they are enabled to bring to birth new possibilities, rather than simply fulfil preset routines, regulations, plans and standards. Too many boxes set into hierarchical 'organograms', linked by rules and rigid pathways, leads to fragmentation and confusion, for everything is in pieces; whereas the meaning – which is derived from a holistic and fluid interaction between situation and group – emerges through the possibility of changing yet consistent relationships determined by the task at hand and the resources brought to it, rather than by generalised procedure.

This will only work to the extent that the 'social competence' of participants is cultivated and respected. Social competence resides in, and is built through, the *practices* of the organism – not the structures and procedures, but the *way* people meet, what they meet around, and the levels of interdependence which are encouraged. The element of earth, of grounding, therefore, should not be seen as merely (or even primarily) a structuring, but as the development of practices which may help the organism to gain a measure of authority over itself.[2]

This 'elemental' sensibility towards facilitating change is relevant for all social organisms, however small and young, or old and complex. Take, for instance, a social situation which is so young and small that it has not yet cohered into anything – say, a meeting which has been convened to solve a problem. Is there an atmosphere of warmth and trust in the room, of intimacy and involvement, or is it cold and formal, suspicious and distant? What can we do to move from one to the other? Does everyone have the same information, and can everyone see that information from each other's perspectives? How can we help to ensure that participants understand each other, the group, and the problem at hand, in at least a consensual if not identical way? Is the group jumping to solutions before it has had time to adequately understand the heart of the issue at hand, not realising that adequate time spent on understanding (the what) will enable the solution (the how) to emerge out of the dynamic of the situation itself? When final decisions are taken, are these apparent decisions or real decisions; how may they be grounded? Thus can this approach begin to guide appropriate intervention.

At the other extreme is the complex bureaucracy, hidebound and very set in its ways, employing a vast range of people from different walks of life, unconcerned with living process and over-concerned with procedure and regulation. Here, too, this approach guides appropriate intervention, though we cannot expect simple compliance by the system or by the process of change itself. The process will be uneven, it will stutter and stagger and fall and get up limping, it will fizzle and pop and tease and tempt and leave you in the lurch just when you thought you were through, and then pick up again just as you were about to abandon the organism completely. It will power ahead in some departments while lagging in others; some people will get it and be inspired, others will get it and hit a new level of despair when they realise how far their colleagues have to go; while yet others may get only a snippet and embrace a new way of working which misses the point entirely.

The four 'phases' which form this approach can be seen as rites of passage. Through the very daunting challenge of metamorphosis there are distinct stages of initiation which must be undertaken, through which the organism must pass, one way or the other. The approach can be used to 'see' the narrative thread of the intervention process itself; we need to maintain a sense for the whole of the intervention even as we work in the parts.

EXERCISE

The Elements

An appreciation for the different phases of intervention and metamorphosis is cultivated through increasing intimacy with the elements themselves. Developing such sensibility is a joy, as if an acquaintanceship of many years had suddenly blossomed into a friendship, and depths are revealed in the other which one had never even intimated. In this case, the other is the world around, opening its secret self perhaps for the first time.

We tend to clutter our lives and interpose all that clutter between ourselves and the essential elements which build our world. And all this wealth of artifact and assistance takes us further from the spirit which infuses that world. It takes great effort to simplify sufficiently to enable the elements to emerge.

Try to find the time to sit quietly beside each element, in gentle communion. Sit before a campfire, watching the flames rise and the embers glow, feeling the warmth, bathing in the light emitted, smelling the smoke as it billows. Feel the rush of the wind, the gentleness of a breeze; watch the clouds drift, feel the air on your skin, the scents tickling your nose. Sit beside a flowing river, watch the flow, the unconscious artistry of the sculptor's hand in the river; stand in the rain, enter the rainbow formed in a glistening dewdrop as it trembles on a leaf before the rising sun. Walk, stride, feel the earth beneath your feet, the weight and tread of it; feel the depths of it below, and the firm and sculpted forms above.

Observe the play of elements as process manifesting through the conduit of a plant. How warmth, air, water and mineral all work together. How the plant anchors itself firmly in earth while reaching up into air, and how it transforms both mineral and air; how it draws water – and allows water to be drawn – through itself. How the sun rules its metamorphoses through providing and withholding light and warmth.

Using the abstract drawing exercise in 'Drawing', pp. 36–7, try to represent the inner gestures of what you are seeing and experiencing through the use of line only; try to characterise and depict the elements in this way. Enter into them through your fingertips. Then use colour, still in abstract mode, where colour itself enables form to emerge, but signs and symbols and representations are not used at all. Do not plan such drawings, but try to allow your inner sensibility to express itself through colour and line, by entering, almost meditatively, the space of the element.

Once you have come this far (or if not through drawing, then through focused visualisation) begin to allow the elements to metamorphose into each other. Observe how the darting of fire becomes the flow of air, how the lightness of air becomes the denser liquidity of water, how this further densifies until the solidity of earth is precipitated out. Work backwards again. Note how the form of earth and body and physical organ – like ear and heart – bears the imprint of the original flow of water; how water itself etherealises into vapour and air; how the origin of air's movement shimmers in the heat which rises off fire, the primordial originator.

Characterise the elements through the use of metaphor. Attempt to capture their transformation one into the other, and their infusing of all nature, through the writing of poetry, or through short prose passages. In this way, think about the characteristic of each element; then take someone you know, and see if you can experience them in the light of the elements – which element is strongest in them, which do they hardly carry at all? Think into the elements as the basis of the four temperaments: how fire translates into the initiating energy of the choleric, air into the pervading and mobile interest of the sanguine, water into the ease and centredness of the phlegmatic, earth into the grounded conscience and integrity of the melancholic. How too their shadows arise: the dominance of the choleric, the flightiness of the sanguine, the comfortable compromise of the phlegmatic, the relentless doubting of the melancholic.

Take an individual or a group, live with them in your thoughts, write about them, see how the elements live within them. Think of the elements within yourself, what dominates and what recedes, what light and shadow is cast thereby, what could you do to further balance yourself. Think too of your life as a practitioner, and the elemental and temperamental characteristics you bring to your practice; use peers to reflect with in this way, so that you help each other to move through your blockages and resistances, and through that which has become so ingrained that it comprises the very way you see.

Immersing ourselves in the elements in this way, they become the medium within which we swim, and gradually we are able to recognise – with an ease and an immediacy – the particular ways in which they may be working through the process of a specific situation, or the process of an intervention itself.

19
Consolidating Change

Firm and yielding transform each other.
They cannot be confined within a rule;
it is only change that is at work here.

I Ching

The dynamics and mysterious movements of change 'cannot be confined within a rule'. Yet how we love our rules and measures, our yardsticks, our criteria and standards, which, we believe, tell us what is happening, and free us of the rigorous demands of consciousness. More and more, as short term productivity and satisfaction become the coin of a globalised realm, as we persist in the illusion that social life can be planned for and controlled as if it were a simple and unliving object, we demand criteria by which we will judge whether our social interventions have been effective or not. Management wants to see the changes written in 'bottom lines' of quantity and productivity; the development sector, with a much broader brief incorporating such concepts as equity, civil society, poverty reduction and human rights, would also like to see changes measured in quantitative terms collectively agreed upon. And the emphasis is always short term, for how else can we monitor progress, how else can we make choices regarding the best use for the money we would spend? How to assess the value of a social development practitioner, or practice?

The problem becomes the more profound when we acknowledge that any agreement between social practitioner and social organism must be clear and transparent, or in the confusion of the change process, given that the practitioner is both inside and outside, both responsible and at the same time merely a guide who cannot force the development process, there can arise conflict and contention when things do not go 'as expected'. Yet how to set measurable objectives when the entire point of a development intervention is that it is intended to enlarge the organism, to open it up, and that therefore the results of the intervention cannot quite be predicted beforehand? Indeed, those that are predicted and achieved are often the least relevant, and the most misleading, for they must inevitably

156

relate more to the intervention (output) than to the impact or actual outcome.

We have to learn to assess change differently. If we refer to the development of a social organism or of the individuals within it, or if we refer to the resolution or transformation of a social situation, we can only assess whether real change has occurred through paying attention to the pattern of relationships which emerge, often very gradually, from out of the situation or intervention. Any social situation – or organism, or whole – is a result of a particular pattern of relationships, and any change will see a change in that pattern. There is simply no other, and no better, way to assess change than by looking to the relationships at work in the situation, and at how they are (or are not) transforming.

In the final analysis the marks of development are awareness, ability, robustness and resilience, foresight, confidence, compassion, creativity, self-criticism, and sheer tenacity. They are measured through relationship to self and other. Infrastructure, economic opportunity, rules and policies and structures, committees and training courses and dams and roads and markets – all these may help or hinder the development of these qualities of relationship, but they cannot substitute for them. And, once again in the final analysis, all these will form as the result of such qualities of relationship, not as their cause. The dynamics are complex, and economic opportunity, or structural change, can certainly affect quality of relationship; the actual details of social interventions are endlessly variable and depend on the particular and unique situation being responded to. But the measurement of whether real change and development has occurred, or is in process, must focus unswervingly on quality of relationship.

If we are to assess movement and change, then we must assess relationship, and particularly changing relationship. If we are to assess the efficacy of interventions, then we have to read myriad relationships which cannot be separated out from each other as distinct cause or effect; and we have to recognise ourselves as being an integral aspect of the relationships which form – therefore also often in process of change. This means that we have to assess the invisible, for we have already seen that relationships are the intangible fields of connections between 'things'; they are not physical objects but emergent wholes which lie behind and between and enfolded in the material world of quantity and object. They cannot be seen in the conventional way, but only through an intuitive appreciation and receptivity which enables *quality* to emerge, to reveal itself.

Take the remark raised in Chapter 13, attributed to Miha Pogacnik: 'Mastery does not mean having a plan for the whole, but having an awareness of the whole.'[1] If we were to assess the development of an individual or group by whether it had a plan or not, then our task would be confined to the investigative techniques required for evaluating quantity, weight and measure, tangible product. But how can we measure 'awareness', and how can we ascertain the precise role our intervention might have played in inculcating it, when so many other things have happened to the organism in the same time?

For the understanding of social development, for the appreciation of relationship and social change and the role played by specific interventions, a different way of seeing is called for. In the words of one researcher who was investigating the effects of wilderness leadership programmes on 'Youth at Risk':

> My approach has been receptive, listening to and recording the stories told by young people and elders. The term 'research' conjures up the stuff of high academia ... a task only some can carry out. This is not so. I have drawn from the stuff of humility and interest, the humility to sit quietly and listen, and the interest which motivates the kind of questions and enquiry which encourages people to tell their story. These qualities are inherent in all of us. Listening to and capturing the stories of our people ... is of great value and is accessible to anyone wishing to do this work.[2]

We cannot penetrate social change through the search for explanation and for quantifiable indicators. These will help, but they will not tell the full story, and they might even mislead us. We must intuit with the whole of our being, become intimate with the phenomena, develop a sensibility for them. Techniques and tools such as questionnaires and graphs will elicit useful information but may also obscure through distancing us from the living warmth of the system as it unfolds. The tendency to denigrate our powers of judgement in favour of instruments and tools holds great danger; not least, it avoids the necessary task of developing those powers of judgement. In so doing it forces the very practice of social development into hiding, and renders us less able to critically assess and adapt our relationships to the organisms we work with as they develop. As Goethe notes: 'Number and measurement in all their baldness destroy form and banish the spirit of living contemplation.'[3]

And Goethe once again:

> The process of measuring is a coarse one, and extremely imperfect when applied to a living object. A living thing cannot be measured by something external to itself; if it must be measured, it must provide its own gauge. This gauge, however, is highly spiritual ...[4]

The phenomenon must be enabled to reveal itself, for it does not need to be explained by reducing it to something else – understanding it will allow it to be revealed in its essential wholeness. Thus the use of story, of narrative, of imaginative exploration. To characterise the phenomenon under discussion, to enable its character to emerge, we employ a mixture of detailed observation – plunging into the phenomenon, paying attention and being receptive, and imagination – the use of metaphor, symbol, image and myth – for the invisible speaks through these things in order to become 'visible'.

A living qualitative reading will never deliver the kind of information required by the managers of 'academic' research, finance and bureaucratic procedure, but a social development practice will have to live with this, subvert these tendencies and thrive on a new form of struggle. To quote Goethe once more:

> Above all we must remember that nothing that exists or comes into being, lasts or passes, can be thought of as entirely isolated, entirely unadulterated. One thing is always permeated, accompanied, covered or enveloped by another; it produces effects and endures them. And when so many things work through one another, where are we to find the insight to discover what governs and what serves, what leads the way and what follows?[5]

Where indeed? Goethe himself provides our answer:

> In so far as we make use of our healthy senses, the human being is the most powerful and exact scientific instrument possible ... the human being stands so high that what otherwise defies portrayal is portrayed in us. What is a string and all mechanical subdivisions of it compared with the ear of the musician?[6]

We have a choice, as social development practitioners – to be technicians or artists, or strive to be both. Each will reveal some aspects of change and deny others. But if the whole is to be held and respected as living process then some form of artistry is requisite. We will not understand until we trust ourselves enough, and quieten ourselves enough, to hear the silent melodies of spirit playing on the instruments of matter. When we focus on external and quantifiable indicators we miss the point, the melody, the whole. Numbers and graphs and tables are no substitute for an intelligent reading convincingly conveyed.

One cannot eschew the use of indicators – it is necessary to be able to say what one is aiming for, and to ascertain how close one has come. But they should be used with a certain timidity, for we cannot measure relationship, and we cannot measure change; we can only appreciate them. Indicators may aid such appreciation, but they often tend to overwhelm it. They are not the point – change is the point, and we have to be sure that we are looking at the process of development in each instance and at all times, rather than ticking off a list of (either quantifiable or inevitably ambiguous) indicators.

We need to enter the situation and tell the story from the inside, out; we cannot presume to see if we construct independent indicators and then use them to judge from the outside, in. When looking from the outside in, there is very little that is required of us; we gather data, apply criteria, and judge – we are not necessarily affected at all. This is analytical consciousness, onlooker consciousness, at work. But measurement affects what it purports to measure, and a reliance on external indicators distorts what is trying to express itself. What we are looking at becomes diminished, limited to our preconceived possibilities. Development is about moving beyond the boundaries, instead of limiting potential. When we begin to think as participants, and apply holistic thinking, we learn to read in an entirely different manner, and every nuance of the world affects and is affected by us. We are *in* relationship while we are assessing relationship. Everything matters, our inner as much as our outer reality. We cannot pre-plan; we can only prepare ourselves, and so allow whatever is struggling to manifest, to emerge into wholeness.

As development practitioners, we must plunge ourselves into the ongoing story of the social organism as it is being lived, make sense of it as it unfolds, and build within ourselves sufficient depth of resource that we may be prepared to offer an appropriate and responsive intervention when necessary. Rather than rigid planning and the assembling of tools and techniques, what is required of the competent social practitioner is rigorous preparation, and the building of surplus inner resources.

This does not mean that we work in the dark, unaccountable and unable to assess where we are. The field of change is not so obtuse as all that, though it will seem so to those who are distant from the field – often the bureaucrats, the managers, the trainers and technicians, the desk and computer people. The more distant one is, the more one clings to the fabricated world of indicators, the illusion of onlooker objectivity.

The problem with 'measuring development' is that measuring implies distance and objectivity; you measure something 'out there' against an 'indicator' which is independent of the phenomenon, and you yourself are unaffected. We can indeed appreciate and assess the process of an organism's development, and the forces which influence it, but not through such 'measurement'. Only from the inside, out, as it were. Only by entering into intimate relationship with the situation (thereby also becoming part of the situation) will we be graced with enabling the situation to speak to us – because unless it speaks, we will only hear the clatter of our own counting machine. Each situation is unique, and each situation will only speak to those who are really paying attention, and who are prepared to be moved.

Intimacy is the hallmark of a developmental practice. It is the essence of being in relationship. These considerations are fundamental to 'reading development'. When we are not prepared to risk relationship then we protect our space by measuring others. This can never render an accurate reading; neither can it be called developmental.

EXERCISE

Working with Peers

The previous chapter presented us with the clear responsibility of honing our own abilities to read, rather than relying on the limited value of external assessments and evaluators. There is no better way to do this than the ongoing rubbing up of our individual practice against the assessments and challenges of our peers, and against the standards of their own practices, so that through the interaction with peers each practitioner grows both themselves and those around them.

There are many exercises through which peers may challenge, support and develop each other – only a few of these are noted briefly in what follows. But the individual exercise possibilities are less important than the necessity to set time aside to engage in such collaborative learning processes. Time which is not interfered with by business issues, questions of policy or structures for meeting and organising, telephone calls and the many work-related issues which beset our day. These must be times for learning and for unlearning, through reflection on practice. For nothing else. Creating such uncluttered, clean space takes deep commitment. It takes, too, high levels of trust and confidence and nurturing; one is putting one's practice on the line. We can easily claim great things for ourselves when the conversation is kept general; but here, the idea is to make it as specific as possible, so that 'on-the-ground-practice' is exposed and considered deeply by all peers involved.

In the first place, one-on-one sessions. I refer here to some kind of combination of supervision/mentoring/coaching/developmental counselling. There is much that has been written on these various methodologies, some of which is referenced in the notes. The detail is not important here, but the fact of structuring such meetings and times is. In a regular rhythm. Written reports will help. These sessions do not have to be undertaken within the patterns of organisational hierarchies; to the contrary, I am concerned here with peer collaboration, rather than with managerial processes. Peers can take it in turns to mentor and be mentored, so long as the relationship and shifting roles are structured and conscious. One-on-one sessions can also take place in a very different guise, where practitioners collaborate on actual work done, and feed back to each other their observations of the other's practice as the work proceeds. This is invaluable, though not easy to arrange.

The rest of these practices involve working with peers in groups. As a start, the sharing of reflective reports, which strengthens a number of faculties. Reflective reports are relatively brief – say two pages – reflections on a particular topic (which all the group are addressing themselves to), or on a particular case that you may be working on, or on patterns that you are discovering in your work, or on yourself as practitioner over the last while, and so on. The group meets regularly and agrees on the topic for the next reflective report. When time comes to meet these are then circulated and read, and the group deals with each one, either singly or collectively. The idea is for the writer to test his or her ability with respect to creating articulate and succinct pictures which really do capture an authentic essence of what is being described; the attempt should be to present a living whole. The others then try to characterise the picture presented, to discover that essence, before adding or challenging or questioning. What new questions are raised, for the individual and for the group, by the reports and the resultant discussions? In this way, each time, provided there is real conversation that takes place, everyone's level of practice is raised at least one notch. This kind of collaborative work thrives on real conversation. More than we usually understand by discussion or dialogue, we try as a group to build a picture through the interweaving of our thoughts and words. The quality of conversation depends on us all; not just what we say, but how we listen, how we respond, how we bring questions to bear, how we place our next thought.

The use of case studies is central to practitioner development. I do not refer here to the use of case-study material gleaned from books and training manuals (which may be helpful in themselves) but rather to case studies which are presented fresh by the practitioner whose case it is, to a group of peers who will listen to the story and respond, thereby testing both themselves and the presenter. Through the process everyone will learn. There are a number of ways of dealing with such case studies. Here follow a few.

The presenter presents a case which has reached a certain point of resolution, either of success or failure. The case is presented as a story in which only the facts of the presenting situation, the intervention(s), and the process of unfoldment are told. The presenter is constrained not to talk about his or her reasons for various interventions, the principles or guidelines or values being worked out of, and so on. All of this is omitted. The group, either as individuals or pairs or in small groups, having listened, is asked to respond by characterising the presenter's practice. What must be the nature of that practice, what its reasoning and underlying values and principles and guidelines, for it to result in the kinds of interventions which occurred? And how does this practice judge success or failure, what are its inherent criteria? Following such feedback, a more general discussion follows, ending with the presenter summarising what they take away from the session, and possibly with everyone mentioning what new questions they take away with them (see 'Questioning', pp. 183–4). Through this process the presenter learns a great deal about what his or her practice is really about, and gets experience in succinct and vivid presentations. The rest of the participants hone their faculties of observation and thinking through to the essence, their abilities to listen without projection, and their understanding of their own practice.

A different form of case study may be used when the presenter has not reached resolution in the process. There may be doubt, confusion, questions as to the next step. There may be an idea of what the next step might be, but no certainty. The practitioner's observation may be faulty, or their understanding of the situation, or their choice of intervention. Something is stuck. The practitioner presents the case up to this point. Individuals or small groups then go off and work with the case, bringing back their reading and the process that they would follow. These are all compared, everyone learns from others' practices and faculties and areas of particular expertise, and the presenter is helped with the real problem they have brought. A variation, where a case has progressed through several crises or periods of resolution or significant shifts, is to present the story up to such a point, where an intervention decision must be made. Groups are then asked to go off and work out their chosen intervention, as above. After sharing these, the presenter describes what really happened, and proceeds with the story to the next point, where the same process occurs. In this way the real (long-term) process of development and developmental facilitation is entered into by the group of peers.

Finally, a case – unresolved, or with difficulties, or needing assistance, as above – is presented by someone in front of a group of peers, but in this case one of those peers acts as 'consultant' to the presenter. The consultant is there to help the presenter find clarity, understanding and resolution; to help the presenter to develop his or her practice. So while the rest of the group listens, a conversation takes place between presenter and consultant. When it is over, the group feeds back, not to the presenter on his or her practice and story (though this may have very interesting angles which the group can go into later as a separate part of the process) but rather to the one who acted as consultant; and the feedback concerns the consultant's development practice, which the group have now had the opportunity of observing first hand. In this case the group has access not only to the consultant's practice but to the presenter's response (that is, to the effect that the practice has had on the 'client'. The client in this case is not a social organism, but the group is nevertheless able to improve its faculties of evaluation and assessment).

If all the exercises mentioned in this section are performed by a group of peers coming from the same organisation, then there is the immeasurable added advantage of a common practice being forged.

Part IV

Practice

Nature ever flows, stands never still. Motion or change is her mode of existence. The poetic eye sees in Man the Brother of the River and in Woman the Sister of the River. Their life is always transitions. Hard blockheads only drive nails all the time; forever ... fixing. Heroes do not fix, but flow, bend forward ever and invent a resource for every moment.

Ralph Waldo Emerson

Whoever uses the spirit that is in him creatively is an artist. To make living itself an art, that is the goal.

Henry Miller

Please remember ... it is what you are that heals, not what you know.

Carl Gustav Jung

20
Discerning

As an analyst, I am always aware that something inside that person knows.

James Hollis

All that has gone before implies that this alternative way of seeing, which is able to intuit the invisible whole, can be accomplished with a fair degree of accuracy. Yet how to acknowledge that we are able to see the 'truth' of a situation, today, when everything has become relative? The grand narratives of the past, which used to provide us with meaning, have been shattered and fragmented. We have been taught to trust no longer the pronouncements of those who lay claim to the truth; we are told that all we can be certain of is the matter which we can see and touch (and even this has become suspect). We have attained a degree of freedom from the tyranny of tradition, culture and dogma which we will not take kindly to losing again. Who is to say what is true, and what is false?

There are many who have attested to the existence of a spiritual world which exists beyond the material world in which we live and die. Such a belief in a spiritual world has allowed and enabled the search for truth; it has testified that, however misguided our attempts – and perhaps we are predestined to continued failure because the quest for meaning is more important than any answers we might reach – there is a truth, or truths, an objective reality, for which it is legitimate to explore. The journey has led us to search within ourselves, and without; there are many who have laid claim to such truth, and developed 'codes of conduct' based on such assumption. Often these have turned out to be as constraining and oppressive as they have been beneficial and liberating.

With the progress of our material civilisation many of these assertions have been held up to ridicule: God is dead, there is no objective reality out there, there is no spiritual world, no world beyond the empirical world which we can see and touch. The rise in the hegemony of the natural sciences has both aided and been abetted by this point of view – that there is no reality beyond the physical, or material; that life is a result of chemical coincidence, and that there is no ultimate meaning for which to search, nothing beyond the prosaic functioning of a material world.

As we enter the new millennium, the battle lines between technical mastery and spiritual intimacy are drawn more starkly than ever. The new sciences have revealed a world which does not resemble the world of mechanical and discrete certainty which preceded them. Everything is relative, ambiguous and contradictory: waves can be particles and particles waves, simultaneously, and depending on how the observing is being done and who is doing it. Nothing is any longer quite as it seems. For some, this is a further indication that there are no ultimate truths, that the best we can hope for are temporal approximations, that the world we observe is a creation of our own imaginations. For others, the move away from certainty and dualism, from a world of discrete things to a world composed entirely of relationships, ever in flux – these discoveries have opened up new vistas in which the world of the spirit appears to have more reality than the world of matter, largely because there is, it seems, no longer any matter to speak of, but only patterns, forces, relationships. Thus for some the new sciences represent further proof that the world is coincidental and ephemeral, and that there cannot be any ultimate truth; for others, the same 'facts' point to a world of spirit which is more real than the world of matter. The only thing which is certain, and this has been said often enough now, is that there is no certainty, and that everything is open to interpretation.

It cannot be coincidental that the new sciences have grown up alongside profound shifts in the world of social relations, politics and economics. Here, democracy has become the formative and normative value by which to judge action. And democracy means equality on all levels – at least theoretically; it stands for pluralism and diversity and collective decision-making. If there is no certainty, then it makes sense to call on the will of the majority, which becomes the final arbiter. Here lies a degree of freedom from tradition, hierarchy, and elite minorities which is indeed profoundly new.

Yet it has also brought discomfort in its wake. The very lack of a final arbiter – in the form of either tradition or hierarchy – means that there is no centre, no overarching value or values from which to draw meaning, and out of which to make choices. We are free to make choices, but have little to base them on. And into this vacuum has arisen a culture in which money has become the dominant value, and those with money the dominant power. Thus along with democracy – and now we have no alternative power or protector to which to turn – has come a virulent form of materialism which demands slavish adherance simply for the sake of survival; non-economic (or social) values have largely been swept aside.

We are all free to pursue wealth and happiness. Yet a new totalitarianism has arisen, an insidious one against which we cannot fight, for there is no one in charge. There is no centre. We are all of us caught in the twin pincers of elite globalisation and unfettered capitalism and there is nowhere to turn. In the name of democracy, we are in charge, yet none of us, individuals or governments, really control anything. There is no central power anymore, yet we are all powerless. There is no common morality, no commonly recognised ethic, no cultural tradition to which to turn; the only power, the ultimate value, is money. Democracy seems to have led us to a point where we are all equally helpless in the face of a sort of material anonymity. There is no truth, no spirit; we are free to pursue anything we wish, but it is an eerie freedom, a void. It has become a journey without destination, without the possibility of destination; we are, it seems, adrift.

This leaves us in a very hard place. I speak of these things because they profoundly influence the work of the social development practitioner. If there is no possibility of seeing through, with a fair degree of accuracy, to the reality of the non-material (spiritual) patterns which form the whole; if there is no possibility of accurately apprehending the narrative thread – if it is all a matter of subjective conjecture, possibly not even real – then where do we turn?

More to the point, the individuals who people these organisms with which we work are placed in the same situation that we are; they are also faced with making choices while being told that there is no foundation upon which to base those choices, that truth and objectivity are a myth, that compromise and speed and technical expedience are more relevant than value and principles. If we cannot see, then we are all caught in a blank place of moral paralysis. Without some approximation to a truth, an objective reality, which lies beyond the material, then – to quote Fritz Glasl, following Karl Popper – 'in the last instance it is paradoxically violence ... which remains the means to impose one's will on other people'.[1]

There must be a way through this dilemma. There is no doubt that we can no longer simply depend on the views of those others who claim to have seen beyond the material, and who bring their spiritual insights back to us in the form of learning which too quickly turns to dogma. We must indeed be wary of those who claim to have seen, when the world is so relative, and there are so many different perspectives, depending on where one stands. In the absence of a given and common centre we must be able to find our own centre. We have to be able to see for ourselves, right through to the underlying spirit which animates and which is so 'invisible'; we must be able to find our own truth, or the truth inherent in a situation at a given point, we must be able to discern the whole which is the real meaning of a situation, to penetrate through to the narrative thread – else we are condemned to work in the dark, with no real means to discriminate, in a cold and lifeless world of profound anonymity, where we have no more significance (and probably far less) than the dot in '.com'.

There is in fact a way through: it is to remember that the laws of spirit reverse the laws of matter, and that life has a logic and dynamic of its own. We can move beyond the impasse caused by 'either/or', and move into the realm of 'both/and'.

Consider this: in the hills of this wilderness realm where I sit and write, there are, invading the pristine fynbos-covered slopes charac-teristic of the southern cape of South Africa, thousands of alien trees called Silky Hakea, originally brought in from Australia, which are now taking over this entire biome. One of our tasks is to eradicate these very ugly, ravenous trees before they eradicate the fynbos, as they will if left alone. A number of us spend time on the hillsides, therefore, cutting them down, trying desperately to contain the spread. The remarkable thing is this: when we first approach a hillside, with these trees dotted amongst the fynbos (which is a low-lying bush), we simply observe the blot on the landscape and set to work. As we work, however, a wonderful thing happens. It feels that we are releasing light – that these trees, so beautiful in their native Australia, create a palpable darkness over here, as if they have pinned shadows down where they stand, shadows which tower over the landscape and slightly diminish the glory of the fynbos itself. As we work, more and more light is released, until, when we look back on a hillside which has been entirely cleared, the brightness hurts the eyes, taking on the proportions of the miraculous. Light literally streams forth.

One could say that this is simply conjecture – knowing the havoc these trees cause, we are simply pleased that they are gone, and make assumptions about something that we cannot really see, and cannot even say exists – something invisible, something of a spiritual nature.

But we have all seen it, all remarked on it, all felt it and been moved by it in the same way. We have seen the 'invisible' shadows, and seen the 'invisible' light which has been released. It is not the tree, nor the hillside, nor the fynbos; it is the relationship between these things which manifests in darkness, and later in light. These trees cause no darkness in Australia.

It is all relative then, it depends on circumstance, but this does not mean that real relationships, objectively real patterns, and the wholes which emerge from them, cannot be directly seen, observed, intuited. Silky Hakea are not inherently bad, nor dark; fynbos is not inherently good – but the relationship between them is a real though 'invisible' pattern which can be read, with accuracy and certainty. There are truths out there, not all of them eternal. Some we have always known, some arise temporarily, and are only relevant for specific situations.

The point is that we can observe real pictures, real wholes which are no-thing. That we can be mistaken about them does not mean that they do not arise, or that we have no right to look for them. It might mean that we must improve our powers of observation, or simply try again. It might mean that we must collaborate more closely with others in order to obtain a clearer picture, and in so doing hone our own ability. But we must try if we are to work with the social with any degree of depth or authenticity.

The approach of 'both/and' can be put like this: we can see through to what is real, so long as we realise that what is real is always changing, is always in a process of metamorphosis; and moreover that we affect the reality of what we see even as we apprehend it, so that we are never separate from the truth we perceive. With this understanding, we can legitimately claim to discern what is actually happening, though we cannot (conventionally) 'see' and touch it, because it arises in the spaces between the things we see and touch, rather than in those things.

This is a far cry from received truth, from the dictates of religion or tradition, the pronouncements of demagogues, or belief or dogma. It is, in fact, the very contrary. It is not a truth in itself, but a way of perceiving which allows the world to reveal itself, through its changes and unfolding development. It is a way of seeing which does not dismiss any aspect of the world, but rather permits the world to reveal itself in all its manifold and contradictory aspects. It is a way of seeing which enables people to find their own truths, rather than simply reach expedient compromises. It is a way of seeing which enables the character of situations to reveal themselves. It depends on trusting your perception while knowing that you may be wrong, of developing yourself as a sense organ while at the same time recognising that you are a participant in the drama you perceive. It is the practice of discernment, of accurate intuition. It does not deny, but embraces relativity. At the same time, it does not deny or avoid necessity.

Rudolf Steiner, in a wonderful little book called *Practical Training in Thought*, notes that true practice in seeing presupposes a right attitude and proper feeling for seeing. He approaches the task through trying to inculcate a right approach to thinking. We must say to ourselves, he notes, that: 'If I can formulate thoughts about things, and learn to understand them through thinking, then these things themselves must first have contained these thoughts. The things must have been built up according to these thoughts, and only because this is so can I in turn extract these thoughts from the things.'[2] Thus we begin to experience the 'thing' as alive and growing.

He goes on to say (bear in mind he was writing in the early years of the twentieth century):

> It can be imagined that this world outside and around us may be regarded in the same way as a watch. The comparison between the human organism and a watch is often used, but those who make it frequently forget the most important point. They forget the watchmaker. The fact must be kept clearly in mind that the wheels have not united and fitted themselves together of their own accord and thus made the watch 'go', but that first there was the watchmaker who put the different parts of the watch together. The watchmaker must never be forgotten. Through thoughts the watch has come into existence. The thoughts have flowed, as it were, into the watch, into the thing ... Thus when a man thinks about things he ... rethinks what is already in them. The belief that the world has been created by thought and is still ceaselessly being created in this manner is the belief that can alone fructify the actual inner practice of thought ... When a person feels the full truth of these words, it will be easy for him to dispense with abstract thought.[3]

To dispense with abstract thought is to put aside intellectual, analytic thinking, and to adopt a more holistic, intuitive consciousness, one which is able directly to see the 'thoughts' which infuse and build the systems with which we work. We try to look at things not as finished products but as 'coming-into-being'.[4] Practising this, the thoughts which created (or are creating) the thing become articulated as relationships which, combined, result in what one sees. If one begins really to look at things in this way, the world gradually comes alive, and one sees things less as a given set of discrete physical objects and more as a fluid, mingling 'coherence' of developing life processes, which have more the consistency of water or smoke than they do of metal or brick. One begins to see the relationships between things, and the repetition of these relationships which leads to patterns which provide the underlying form to the things which surround us, to the systems with which we may be working. This is impossible to quite articulate in the linear patterns of language. It needs to be experienced, and practitioners can only learn to experience these things for themselves. Experience can never be granted, it can only be discovered.

Christopher Alexander, an architect who has developed a similar way of seeing for the task of designing buildings and towns so that their patterns form a whole which is healthy and nurturing of life, says this about the process of seeing accurately:

> though this method is precise, it cannot be used mechanically ... Indeed it turns out, in the end, that what this method does is simply free us from all method ... The more we learn to use this method, the more we find that what it does is not so much to teach us processes which we did not know before, but rather opens up a process in us, which was part of us already.[5]

Here is the key to developing holistic seeing. To open ourselves up to what we already know, but are afraid to use; not to painstakingly learn a new technique, but to engage in exercises to move beyond technique. Honing our humanity. Not to impose a method on the world and on ourselves, but to become intimate with the world, as it unfolds, so that it speaks to us. The power to do so lies within, but

> the power has been frozen in us; ... we have it, but are afraid to use it; ... we are crippled by our fears, and crippled by the methods and the images which we use to overcome those fears ... *it is not an external method, which can be imposed on things. It is instead a process which lies deep in us; and only needs to be released.*[6]

Initially self-conscious discipline must do its work, but we learn the discipline in order to shed it, to move beyond. To see clearly we have to cleanse ourselves, shed our projections and our fears, and the constraints of convention. It is we who must become whole if we are to discern wholeness in the beings with which we work. To develop ourselves as organs of perception we must work as much on ourselves, on our inner worlds, as we do upon our thinking and our faculties of sensing and observation. Above all we have to loosen, let go, release, in order that the world may enter.

The discerning person can be sensed; the competent social development practitioner can be assessed as much by observing him or her as by trying to ascertain the impact of their work. Eugen Herrigel, in *Zen in the Art of Archery*, points out that his master assessed his progress not by looking at where his arrows were going, but by looking at him as he shot them.

> I no longer succumbed to the temptation of worrying about my arrows and what happened to them. The Master strengthened me in this attitude still further by never looking at the target, but simply keeping his eye on the archer, as though that gave him the most suitable indication of how the shot had fallen out. On being questioned, he freely admitted this was so, and I was able to prove for myself again and again that his sureness of judgement in this matter was no whit inferior to the sureness of his arrows.[7]

Learning to discern is something other than learning a new skill, it is of a different order. It is learning to be someone other. Someone who pays attention.

EXERCISE

Creative Thinking (2)

Here is an exercise which will help develop the ability to see relationships and patterns.[8]

Visualise the building you are sitting in. It is made up of various elements, or things; these vary from building to building, but yours may contain some of the following – passage, room, window, entrance, door, roof, gutter, kitchen, and so on. But what is a gutter? Running along the edge of the roof, just below that edge, is a plastic or metal 'half-pipe', the open section facing towards the sky, the scooped 'U' facing downwards. Now take away the roof, take away the wall, turn the half-pipe upside down and leave it lying in the grass. Now it is no longer a gutter, but a piece of debris which you may wish to remove. What made it a gutter was its specific relationships to other 'things', for example the roof.

But is the roof a thing? The roof sits on top of the building, preventing rain from entering. It is made up of many smaller things, all having a specific relationship to each other – perhaps lengths of wood, sheets of plastic, insulation material, nails and screws, tiles and flashing which is placed at the spot where the chimney penetrates the roof. If these 'things' had different relationships to each other, then they would not form a roof but something else; it is in the specific arrangements of the parts to each other that our concept of a roof emerges. And even if the parts had the necessary relationship, but the resultant structure were not sitting on a building, then it would not be a roof. Its 'roofness' depends entirely on relationship, and it is the configuration of relationships which create the entity. The entity is nothing other than these relationships, and the patterns which form through their repetition.

To achieve easy entry into this exercise, try to write a descriptive piece on one of your building's 'parts' without referring to any other part. The task may well become impossible, because the interconnectedness of everything is increasingly revealed. It is in the tension between our tendency to glib separation and the actual impossibility of separation that the value of the exercise lies. The closer you look at any 'thing', the more you will see that what we think of as a thing is in fact a pattern of relationships between this 'thing' and the 'things' in the world around. The 'room' in which you may be sitting is a specific relationship between four 'walls', a 'floor' and a 'ceiling', two 'windows' and a 'door'. When this relationship is repeated across many different situations then we have a pattern emerging, and it is this pattern to which we give the name 'room'. The room is not a thing, as we may like to think; it is a pattern of relationships. The specific relationships will change from room to room, but the pattern will remain, for it is this which enables us to identify it as 'room'. It is entirely insubstantial, if you will.

Look closer, and you realise that even the 'things' which are relating to each other and forming the patterns are not 'things' at all, but themselves patterns of relationships. Consider the following from the architect and writer Christopher Alexander:

> Consider, for example, the pattern we call a door. This pattern is a relationship among the frame, the hinges, and the door itself: and these parts in turn are made of smaller parts: the frame is made of uprights, a crosspiece, and cover mouldings over joints; the door is made of uprights, crosspieces and panels; the hinge is made of leaves and a pin. Yet any one of these things we call its 'parts' are themselves in fact also patterns, each one of which may take an almost infinite variety of shapes and colour and exact size ...
>
> The patterns are not just patterns of relationships, but patterns of relationships among smaller patterns, which themselves have still other patterns hooking them together – and we see finally, that the world is entirely made of all these interhooking, interlocking nonmaterial patterns.[9]

This exercise asks nothing more of you than that you begin regularly to look at the world around you with these thoughts in mind. Take that which is familiar, with which you interact frequently, and consider 'it' in this new way. Paying attention, we may penetrate through to the patterns which configure it. The more we build our ability to discern relationship and pattern in this way, the more discriminating and differentiating – in the best sense of those words – we become.

More to the point, we gradually build the ability to observe the patterns which are really forming the 'things' we see. If we combine such a way of seeing with peripheral vision, we can increase our ability to experience the living relationships around us – our social world reveals itself as an intricate web of relationships repeating themselves into patterns; so much so that the richness and depth of all that surrounds us becomes apparent, almost amplified, and we begin to experience the invisible as more real than the visible.

If we then enter into social situations with this sensibility, organisations will reveal themselves more and more as manifold relationships building into pattern. We may find that there is nothing else. And that it is these dynamics which need be attended to, and cared for.

21
Co-creating

The future enters into us, in order to transform itself in us, long before it happens.

Rainer Maria Rilke

I return to some phrases incorporated in a sentence written in the last chapter: 'we can see through to what is real, so long as we realise that ... we are never separate from the truth we perceive'. As social development practitioners, we have to recognise that there is no observation, no perception – indeed, *nothing out there at all* – which exists independently of ourselves. We are always and indubitably implicate. Not only do we affect what we see; we affect it by the way in which we see it.

Classical science assumed an onlooker consciousness; that we are observers only, detached from the phenomenon itself, at most manipulating it externally. Here though, is something essentially different, something which has rather radical implications. Because we are never separate from what we perceive, because we are implicate and involved, what we perceive will always be a function of what we bring to it, of who we are. If we are, then, to work towards accurate observation, then such capacity requires ongoing development of the social practitioner him or herself. One cannot rely on techniques and tools, one brings oneself to the intervention, in intimate relationship.

There is very little of comfortable linearity here. Cause and effect become interchangeable and intertwined. We are obliged to engage in a practice in order to enable us to develop the capacity to engage in that practice. Goethe pointed to this, adding that the organs of perception we are talking about are not given, waiting there to be activated, or perhaps slightly atrophied because of underutilisation. He claimed that the required organs of perception must be *developed* as *new faculties*, and that the only way to do so is to be active ourselves in their development.[1]

To work as social practitioner in this way, we ourselves have to become our own instruments. This is the alchemical understanding: the world without and the world within are one and the same, they take from each other, give to each other. We are participants in the unfolding and becoming of those with whom we work; it is through them that we unfold and emerge. This is the essence of co-creation. With such a sense of co-responsibility, we can indeed help to develop, to enlarge and make more human, the social fabric which surrounds and nurtures us like the membrane of a womb.

We are not working with a given; we are building our reality as we go. The world, and every system and organism within it, is in a process of becoming, and we are co-creators with respect to that becoming. Everything we are or do has its resounding effect in and on the world around us. We are never without effect. And we are never unaffected. As social practitioners, our practice will be immeasurably affected by this profound, even radical, realisation. The more we see, the more there *is*. Facilitating the development of others cannot be achieved without facilitating our own; and our own development is dependent on the consciousness of the world around us. As social practitioners we are at the forefront of a co-creative endeavour which is resulting in the world of our future.

Rilke, in writing of God, presents an astonishing possibility – that God was not simply the creator, but is in the process of being created, by us. This resonates closely with what we are saying here. He writes, in a volume called *Letters to a Young Poet*:

> Why don't you think of him [God] as the coming one, who has been at hand since eternity, the future one, the final fruit of a tree, with us as its leaves? What is keeping you from hurling his birth into evolving times and from living your life as though it were one painful beautiful day in the history of a great pregnancy? ... By extracting the most possible sweetness out of everything, just as the bees gather honey, we thus build him.[2]

The same theme is explored in the medieval legend of the Holy Grail. The Grail is a mystical chalice of salvation, the end of the hero Percival's quest, which is attained when, after many years of wandering and mistaking the path, he finally asks the right question, the question which allows him access to the Grail. The fact that success is contingent on the asking of the right question is itself instructive for the social development practitioner, for right use of questions is the guiding light of the profession. But I raise the story here for another reason: the word 'grail' really means gradual, and the heart of the Grail Mystery is that the vessel was formed gradually through the questioning path, the quest; through the search itself. While the quest was in search of the Grail, when the quest began the Grail did not exist – it came into being gradually through the integrity and trial of the journey towards it. Once again, there is no linearity here, no easy cause and effect; quite simply, that which we do turns out to be that which is done. Although it sometimes seems complex, at heart the concept of co-creation is supremely simple.[3]

The essence of the work of the social development practitioner lies in this – that we connect to the underlying narrative of the social organism as it strives towards wholeness, and in so doing help harness the energy which lives in the meeting between archetypal pattern and unique contribution, so that we are able to unblock blockages and release stuck spaces, in order that living process unfold once more.

We have to imagine that we can create together. Current realities, structures, procedures, capacities – all these are given, and as such they have developed out of past endeavour. They become skeletal, dried out, the cast-off husk of past growth. To move beyond, we have to attach ourselves to an initiating energy, the underlying movement and direction of an organism's life processes. We connect thus to the spiritual energy which is continuously forming the organism, rather than attempt to manipulate structures, procedures, plans, ways of working which have had their day, used up all their juices, as it were. Our task is to seek the new, the formative forces which are beginning to form through new activity of the spirit, and to harness that energy through consciousness, transparency, awareness, understanding, and most of all, courage. To enable the organisation, through our own courage, to find the courage to let go of the past, link itself to its own rising tide, and move on into an unknown future.

When we work with social situations, we are working with beings, living beings, very powerful beings at that. Beings who are often so steeped in the gathering detritus of their past, who are so debilitated by the gathering failures caused by numbing routine and habit, who can be so huge and lumbering and unintelligent – like the giants of mythology – that they have lost all nimbleness, all fleetness of foot, all of the energy of youth and hope. When this happens, however large or small they may be, they sit astride our world with a cobwebbed gloom, straddle our hillsides and cast their copious shadows like Silky Hakea over fynbos. And there are others, not in the least unintelligent and very fleet of foot, full of vim and vigour and overarching intent, who may nevertheless be hell bent on manipulation and expediency, all in the supposed name of a better future and a more flexible response. These may roam our hillsides like brigands, intent only on plunder and personal gain, wearing their hearts on their sleeves and their logos on their shirt pockets.

It is with these beings that we engage when we practise the art of social development. If we can recognise ourselves in the organisms we meet, and commit to genuine participation in their recovery, then together we can begin to create something new. And the aim of such co-creative work is wholeness.

We have used the concept of wholeness very often in this book. Used here it has three distinct shades of meaning, though they all revolve around the same core. First, it refers to meaning, as we have had it earlier – that meaning which is no-thing, which exists beyond the parts, which gives the parts their significance and which lies enfolded in the parts. Second, it refers to the process of becoming, as in growing towards wholeness; the process of becoming more oneself. Third, wholeness refers to a particular quality of living integrity which is impossible to define, yet which we recognise when we see it.

The architect mentioned in the previous chapter, Christopher Alexander, refers to this aspect of wholeness as the 'quality without a name' – we know exactly what it is and will recognise it when we see it, but we cannot define it.[4] Referring to buildings and towns, he notes that:

> the difference between a good building and a bad building is an objective matter. It is the difference between health and sickness, wholeness and dividedness, self-maintenance and self-destruction. In a world which is healthy, alive and self-maintaining, people themselves can be alive and self-creating. In a world which is unwhole and self-destroying, people cannot be alive: they will inevitably themselves be self-destroying, and miserable.[5]

Precisely the same perspective can be taken of social organisms.

In this sense, the quality of wholeness and Alexander's 'quality without a name' are identical. Wholeness is not replicable but unique to a particular situation, because it is alive and takes its shape from an intelligent response to the particular circumstances within which it exists. It is 'a subtle kind of freedom from inner contradiction'.[6] When something is not whole, its parts do not fit harmoniously together, there are inner contradictions which tear it apart. When a system is at war with itself it gives rise to forces which act to tear it down – it becomes unwhole. Systems which are not whole are not sustainable, they have no internal coherence, they cannot reproduce themselves and maintain their balance, move into their own future, acting out of their own internal forces. In such situations, external and imposed force is required – policing, laws which demand and coerce and threaten punishment for transgression, rigid structures and procedures which constrain and contain. When a system needs such restraints, when it does not have the internal and living resources – the spiritual forces – to resolve its own dilemmas and conflicts, then it can be said to be lacking in wholeness.

It may be a landscape which has been so invaded that it now needs concrete supports to prevent further erosion. It may be an organisation in which rules and regulations stifle creativity and integrity. It may be a wider social situation in which companies pay no regard (or taxes) for the environmental havoc they cause. In each such situation, wholeness is missing, and the forces of fragmentation have to be countered from outside of the system itself (though often such counters lack substance, and cracks begin to appear in the wider system itself). Lacking in wholeness, in sustainability through internal balance, the system has lost track of its own life process, its own spiritual trajectory. Our job is to unblock, help find the path once more; to promote harmony between the parts so that the whole becomes functional once more.

Such quality, though, such wholeness, cannot be created in the conventional sense – that is, it cannot be *made*, it can only be *generated*. We are co-creators, not creators. We cannot simply bring into being from out of our individual egos that which we choose to manifest. We can only work *with* the organism in question, tapping into the underlying trend of its life process, enabling perhaps, guiding possibly, unfreezing, challenging and supporting, helping that life pattern to become that which it is capable of becoming, to realise its unique possibilities. We cannot design and plan its future, think out its direction, create its potential. As Alexander notes, the quality without a name cannot be made like this.

[I]f you want to make a living flower, you don't build it physically, with tweezers, cell by cell. You grow it from the seed ... You know that any attempt to build such a complex and delicate thing directly would lead to nothing. The only flowers which men have built directly, piece by piece, are plastic flowers. If you want to make a living flower, there is only one way to do it – you will have to build a seed for the flower and then let *it*, this seed, generate the flower ... the great complexity of an organic system, which is essential to its life, cannot be created from above directly; it can only be generated indirectly.[7]

We cannot impose our will on organisations and communities and hope that this will enable them to flower, to become more whole. We cannot mould them to our own desires. But we can see into their depths, discern the underlying forces which move them, and thus enable the system to take the next step in its own path of development. Within 'their very depths' will be found the equivalent of the seed of the flower mentioned above, the 'narrative thread'; which brings us all the way back to the quote from N.P. Van Wyk Louw which opened this book: 'Ons moet probeer om die verborge groeiplekke van die geestelike lewe raak te sien [We must try to discover the hidden growth places of the spiritual life].' Only through connecting to the spirit which moves the whole will we enable a system to regenerate itself on its journey towards wholeness.

EXERCISE

Questioning

Questions are to the social development practitioner what arrows are to the hunter. Of far more value than answers, they must be well directed to find their mark. They are magical tools, wands with which to unlock the secrets of social situations. The better the question, the more it will reveal, and the more accurate our reading will become.

Answers end the quest; questions stimulate it. Answers close things down, and invariably reduce; a developmental response is to help the organism find the next question which can act as a spur to its becoming. Questions help us to penetrate to the heart of a situation; yet it is paradoxically the ability to accurately see which enables the correct question to emerge. If we are not seeing the whole, we will struggle to find the specifically relevant question, the one which opens up movement towards the future; yet it is through astute and respectful questioning that we open up the whole to be read. It becomes difficult, then, to find the practices which may exercise the faculty.

Actual practice – more than exercises – is key; as well as reflection on that practice. There are no techniques here, no instruments to assist, little to make the path easier. We can, of course, become aware of different kinds of questions, and sensitive to their different uses and meanings. There is a difference between 'how' and 'what' and 'why' questions. Some questions open things up, both for questioner and respondent; others hone in on a particular area and seek specificity, confirmation or negation, challenge or agreement. Some can be used as facilitative techniques; some will be more comment than question. Some questions are questions of knowledge, aimed at the past, while others are questions of choice, aimed at the future. There are questions for content and questions for clarification. All these different uses of questions will depend on circumstance and need. All of them can be differentiated and practised, so that they may be fluidly used.

As with wonder, there is no substitute for the cultivation of a questioning attitude. An attitude of curiosity and interest. A questioning attitude also implies that we have questions about ourselves and our life, that we are never without at least a central question. It is questions which stimulate the developmental movement; without questions we become moribund. Many have indeed ceased to have questions, or have managed to suppress those they once had. Others fear their questions, and deem it a mark of weakness to entertain them. A questioning attitude implies that we do not seek hasty answers, but are prepared to have the questions live inside us and, gradually, as Rilke puts it, allow ourselves to live into the answers. Here is Rilke's approach:

> Try to love the questions themselves, like locked rooms and like books written in a foreign language. Do not now look for the answers. They cannot now be given to you because you could not live them. It is a question of experiencing everything. At present you need to live the

question. Perhaps you will gradually, without even noticing it, find yourself experiencing the answer ...[8]

When working with groups, try to begin and end processes with a lifting out of individuals' questions. Questions which lie at the start of a process give it direction and meaning; they stimulate the developmental path. Lifting questions at the end enables that path to continue, given where people have gotten to, instead of closing it down simply because some things have been realised. Whilst doing this, always ask yourself what your questions are at this time, of the process and of yourself, so that your own affinity with questions grows and deepens.

Take groups of questions – say those of individuals in the group – and try to find the central question which may underlie them. In other words, either yourself or together with the group, try to characterise collections of questions belonging to an individual in such a way that that individual is helped to find his or her central underlying question, the whole which is lending significance to all the parts. Or try to find the one central question which is moving the group, from a collection of individuals' questions.

Take a person that you know, and observe them attentively at particular moments for some time. Then ask yourself: what is his or her question? Do the same for a number of other people, and for groups and social situations. The art of asking questions can become second nature.

Try to help individuals or groups come to resolution about future plans or clarity about current problems only through the use of questions – without permitting yourself any other facilitative intervention. For example, in 'Working with Peers' (pp. 162–4) and 'Listening (2)' (pp. 145–6), where there is the opportunity to act as consultant or facilitator or counsellor. We can practise these exercises sometimes while restricting ourselves to the use of questions as the only intervention permitted.

Above all, question everything, all the time. Take nothing for granted, accept nothing at face value. It is often those questions which seem most stupid and irrelevant that expose the vagueness, the inauthenticity, the unacknowledged contradictions at the heart of a situation, and by so exposing, enable forward movement to take place.

22
Emptying

Emptiness is the track on which the centred person moves.

Tsongkhapa

We are used to thinking that there is something to be done, something to be achieved, and that we have to do it; that what is important is product. We work, in the everyday, with material manifestation; conventionally, it is things which we focus on, rather than the spaces between the things. Our deeds gain our commitment, not our withdrawal into contemplation or profound consideration – the very word 'withdrawal' emphasises the point. Yet what if the very opposite was asked of us?

The Tao Te Ching, that ancient book of such illuminated wisdom, reveres emptiness as a way of being. 'One becomes as a babe' is one way in which it expresses the state of purified spiritual awareness, a way redolent of Christ's injunction to become as little children.[1] These are not moral commandments but descriptions of the way the world works. The Tao goes on to say: 'Let his intelligence comprehend every quarter, but let his knowledge cease.'[2] Emptiness does not imply a state of vacuousness, of vagueness; but it does warn that too much knowledge can cloud, or prefigure, one's vision. In which case there is no true discernment. Instead, it recommends the development of intelligence over knowledge, and the *bringing to bear* of that intelligence. Intelligence is not given, it must be developed and honed, and in the process knowledge will be acquired. But the knowledge is beside the point, and may obscure as much as it clarifies; it is the faculty of intelligent and pure observation which is important, and for this emptiness is required.

Social development practitioners run the risk of learning so much through years of experience and facilitation that we gradually cease to facilitate at all, and become instead the repository of expertise, spending more time inputting into social situations than responding developmentally. At such times an impatience can begin to creep into our work – why can the community not see what is so obvious to us? And so we slide into that danger expressed so eloquently by George Bernard Shaw: 'Reformers mistakenly believe that change can be achieved through brute sanity.' Which is why the Tao Te Ching goes on later to say: 'As to those who have knowledge already, he teaches them the way of non-action.'[3]

The way of non-action is the facilitative way, the way of harmony, of gentleness. Non-action allows and enables the world to evolve according to its own evolutionary processes; it does not impose, or manipulate, or seek to control. It is a way which seeks to enter into phenomena and, through increasing intimacy, to help unfold the potential which lives within. Those who have knowledge already must learn to withhold, create the space from out of which new possibilities may emerge. Non-action does not imply inaction, lack of commitment or laziness; it is an active withholding, from a position of strength, to allow action to emerge from beyond oneself.

To get in touch with what is outside we must find that which moves from within. We often clutter our own perception and our own response with too much knowledge, impulsive action, and the obsessive notion that results and outcomes depend on us. Embracing emptiness as a path, we may discover new affinities, new intimacies, new possibilities for facilitating the development of more humane systems.

We find it so very difficult, held within the moral imperative of achievement as we are, to realise that there is great value in its opposite, in acceptance, which is another face of emptiness. Our obsession with achievement leads to two possible shadows of that very achievement. On the one hand we may lose ourselves – succumb to vainglory, vanity and presumption; we may lose our sense of boundary and limit and respect. On the other hand, if less successful we may become unable to really reach ourselves; our lives may become pinched, small, petty, without meaning or celebration, cold and calculating, unable to move beyond the given. The middle ground between these two shadows is a finding of oneself as one grows; there comes then a recognition that we work always within boundaries but are not necessarily constrained, that authenticity consists of recognising limits whilst still working out of freedom.[4]

To this end, the notion and practice of emptiness is invaluable – we begin to see, embrace, accept, and encourage the world of spirit to move through us, to do its will, to move us. So we become connected to the greater whole without losing autonomy – in fact, such acceptance engenders a growing strength and independence. Not simply I, but the evolving world through me. I become as a vessel for the development process itself to unfold through me, I become as a conduit for blocked energies to move once more. We noted earlier how water always tries to form a sphere, an organic whole, by joining what is divided and uniting it in circulation. Within this cycle of circulation plants play such a role of conduit, and it is the same here for us, with spirit forming the organic whole and ourselves being the conduit, the channel, for such expression. If, of course, we allow this to be. Much of the time we block such possibility through our own excessive cleverness. As the poet Rumi has written: 'Sell your cleverness and buy bewilderment.'[5]

In *Zen in the Art of Archery*, Eugen Herrigel refers to such notion of emptiness as

in fact charged with spiritual awareness and ... therefore also called 'right presence of mind'. This means that the mind ... is present everywhere because it is nowhere attached to any particular place. And it can remain present because, even when related to this or that object, it does not cling to it by reflection and thus lose its original mobility ... it can be open to everything because it is empty ... and its symbol, the empty circle, is not without meaning for him who stands within it.[6]

Here too – as with the understanding of an holistic consciousness – the concepts are difficult to describe from the outside, for we tend to try and grasp them from our 'usual' mindset, which is other and foreign. We cannot really grasp the meaning and value of the symbol of the empty circle except from within; we cannot really understand the power of emptiness and non-attachment for our very tangible work in the world unless we seek the experience itself.

This means that we must step outside of the hurly-burly of the day-to-day, at least from time to time, preferably with a disciplined regularity, so that we may begin to feel the energy of silence and the power of detachment. The Quaker philosopher Douglas Steere wrote:

The rush and pressure of modern life are a form, perhaps the most common form, of its innate violence ... It destroys one's inner capacity for peace. It destroys the fruitfulness of one's work because it kills the roots of inner wisdom which make work fruitful.[7]

And the Native American Ohiyesa, a Santee Dakota physician, has been quoted as saying of traditional Native American culture:

there was only one inevitable duty – the duty of prayer – the daily
recognition of the Unseen and Eternal ... Each soul must meet the
morning sun, the new sweet earth and the Great Silence, alone ....[8]

That 'Great Silence' lies at the heart of our humanity; it provides
our connection to that which lies beyond ourselves. It is solitude
and the sacred, retreat and sanctuary, and is formed through the
ability to simply be, on the other side of doing.

Once again, this is not a moral injunction so much as an attempt
to describe how the world works. As development practitioners we
are beset by many, varied, often conflicting processes happening all
at the same time. There may be one underlying narrative thread at
the heart of any social situation at a particular time, but there are
many stories making up this thread, many streams and currents and
rhythms swirling about one another. Take any community, any
organisation, any group you know and think about the number of
relationships which go to make up the whole. Standing within such
diversity – as social practitioner – we cannot afford to simply
constitute another strand. It is too easy to get swept away, to get
confused, to contribute to the confusion. The ability to detach
oneself, to find a place of emptiness within oneself, is essential if we
would presume to assist the healing process. For we must be able to
be inside of, and appreciate, many processes at the same time, and
we can only do so from that place of emptiness. This has nothing to
do with acquiescence, or subservience. Emptiness is not a weakness
but a strength, and is developed *in the wake* of a strengthening ability
to perceive and intervene, rather than as a replacement. It is not an
absence but a withholding of what is already present.

Such 'withdrawal' into emptiness enables us to *consciously* create
a hollow space, which is a place of power – it is in this sense that we
speak of acceptance rather than acquiescence. The task is to become
inwardly silent and so form a receptive 'shell' (much like an ear)
which can become a resonance-organ for those forces which are not
tangible or immediately visible. In people who have not developed
their powers of thinking, of observation and imagination, silence is
absence. But for those who have developed themselves and who
hold themselves back, such emptiness and silence is a presence
within which the secrets of the social situation may resound.

St Benedict, who many centuries ago developed an Order for a
Christianity which seeks to combine all polarities of life into a
wholeness of community, spoke of similar things. He had no

pretensions towards promoting the miraculous but advocated the ordinary life lived extraordinarily well. As with the Tao Te Ching's observation that intelligence should comprehend every quarter but knowledge cease, he

> tried to discern the needs of his time and respond to them. His reaction was not a negative one, even though he was writing in a period of decadence and disintegration. His concern was less to pronounce on the world situation than to show concretely the correct stance towards it – an attitude of 'interior openness to the world'.[9]

In pursuit of such, his Order asked of its novice monks that they commit themselves to three practices. These commitments may be instructive for us as we listen to, and intervene into, social situations.

The first such commitment is termed the vow of stability. It calls for the practitioner to stand still, firmly planted not in any physical sense but within oneself, not running away from who one is. It recognises that the world is in continual change, that all is fluctuating process, and it asks that we neither deny this nor get swept away by it. Thus to find one's centre, one's sense of self, as a point as still as the eye of a storm. To cultivate an inner constancy from which to look out at the fickle and fast changing world around us.

The second vow appears almost to contradict the first. In Latin it is known as the vow of *conversatio morum*, which translates as the necessity to avoid becoming attached, fixed and rigid. It asks a recognition that life circumstances change and that one has to have the openness and flexibility not to resist but to change along with them. I must always and constantly be ready to grow and change and move on, to let go. Such is the condition of consciousness and development.

What appears at first sight as a contradiction is in fact a complementarity. Living between these two poles forces me to encounter a fundamental tension that I can never expect to escape or evade, and which strengthens me as I stand upright within it. I am asked to hold myself in such a way that I stand firm whilst simultaneously moving on.[10]

The third vow is the vow of obedience. It asks that we find the silent and still place within which allows us to listen to the voice, to the narrative thread, to the underlying patterns of our own life, so that we connect that life to forces larger than us, and live it as a search for meaning, in the understanding that such meaning will become clear to the extent that we allow ourselves to listen for it. To learn to *receive* rather than achieve. In the words of David Steindl-Rast: 'Be so still inside that you can listen at every moment to what life is offering you.'[11]

All three vows form one whole; none is able to be lived without the others; each one implies all three. It is the cultivation of this approach which enables emptiness, and which allows the world of force and pattern to speak to us through the clangorous manifestation of matter – which both masks and reveals the processes with which we work.

Mostly, as consultants, leaders, managers, facilitators, we are expected to fill ourselves with the latest theories, with norms and standards and tools and techniques, so that we may gain control over, and dominate, the systems into which we intervene. Yet true social artistry lies beyond all this clutter. It lies in the heightened awareness that comes with detachment. It lies in the emptiness which enables one to remain centred in the midst of conflicting flows and processes. Insight arises when the strengthened mind resists distraction in favour of a tranquillity which, as with still water, allows sediment to settle and a clarity to take its place.

EXERCISE

Shifting the Pattern

This is a particularly difficult exercise (as the name might indicate).[12] At the end of the day, before bed, in solitude, a 'review of the day's events' is performed. We go back in our mind over the whole day, thinking through all that has occurred. What makes the exercise difficult is that this review is performed backwards, in reverse. We start with the last event, that which occurred just prior to sitting down to this process, and we move backwards through the events of the day until we reach the point of arising in the morning.

If you try the exercise you might experience great resistance, so it is valuable to acknowledge and negotiate with such resistance; for the exercise, even performed less than optimally, is invaluable. You will not initially be able to perform this exercise for the events of the whole day – you will not remember them all, or their sequence; the exercise will feel pointless; the events will become muddled as you begin to see connections which you may wish to explore but which are out of sequence; you may get involved in a particular incident or period of the day to the exclusion of other periods. In which case, stick then with a short period, or with one event, and view this period or event in reverse. Or try to accomplish two, from different times of the day. Or go backwards from where you are until you feel that you cannot anymore. Or begin the review from lunch backwards.

Do not do it every night if this is too intrusive, as it probably will be. Do it once or twice a week to begin, or even, simply, whenever you get the chance. If you just begin, with perseverance you can progress further and further. It requires great powers of concentration, but its value is manifold.

It is in fact a version of Goethe's methodology for developing new faculties of perception, his 'exact sensorial fantasy' (see 'Creative Thinking (1) and 'Creative Thinking (2)', pp. 67–8 and 175–6) performed in reverse because this enhances the method. In order to accomplish it, you must observe accurately every event of your day as it unfolds; you have to pay attention, and move through the day consciously. Then too, you must develop your powers of memory, your ability to commit to memory, and later to draw from its contents. This was the essence of Goethe's method – first to plunge into seeing, and then, by way of objective memory, to inwardly re-create the 'coming-into-being' of (in this case) your day. Thus to bring perception and thinking, sense observation and imagination, together so that we may begin to experience the depth of the whole. But this time, backwards.

Why backwards? All of the faculties necessary are enhanced by working backwards, for you can take nothing for granted. Working forwards, your consciousness through the day – the paying of attention, the committing to memory – is not truly tested, therefore not drawn upon to the same extent, not improved upon. Working forwards (at the end of the day) a laziness may creep in, a sliding forwards from one event to the next without much clarity or differentiation; a blurring.

This is the point of the exercise, but for different reasons. When we work backwards in this way, a freeing up of the conventional occurs. We are forced out of the conventional, we are forced to view events afresh; in particular, we are forced to see from completely new and strange vantage points. We can no longer slide easily from one thing to the next; each event becomes discrete, its own whole. A glibness disappears; our actions no longer follow an easy rationality which we can explain to ourselves because, after all, they made sense enough to us when we were performing them. The supposed link between cause and effect, so taken for granted, is broken. Now, in this reverse review, they stand bald and naked, separated out from assumption and obvious inevitability. We can walk around them, as it were; view them from all sides. They are set free from the normal kind of unconscious acquiescence, and in turn, through this process, our consciousness itself is set free. Or, put differently, we become more conscious.

In this way, we learn to access that higher Self beyond our everyday selves; we empty ourselves of assumption and preconception and a clinging to conventional expectation. We detach ourselves, and, thus empty and open, are able to receive new insight and meaning; we begin to roam free over the events and activities of our day. In so doing we may gradually discover a new level of consciousness which is not attached, which is mobile and interested, filled with wonder, inundated with shades of meaning.

This kind of exercise, adapted, can also be performed in other contexts, and in groups rather than in solitude. So, for example, as in reviewing the process of an intervention, or the events of a particular session. A group can review aspects of its biography in this way, or of a day spent together working at a particular task. All of these adaptations are valuable, and, if skilfully facilitated, will serve to enhance the consciousness of the group.

23
Awakening

[S]lowly you get the ability to pull these supersensible beings in a way to the threshold of the physical world ... to the border of the visible world ... I don't want to create the impression I am clairvoyant and can behold the beings in their supersensible appearance. I can't. But I have indeed brought them to the border of visibility. They have become realities. I mean I began to live with these beings in the same way you live with realities you can see with your eyes.

Bernard Lievegoed

Awakening to see spirit and life stream through the interstices between material things is a developmental leap into a world which feels, for each person who accomplishes it, as if it has been inhabited for the very first time.

Such awakening is not simply something that happens, inevitably and irrevocably, like the arrival of dawn. It is attained through disciplined practice, and is itself a discipline and practice. Ever the poet of transformation and development, Rilke reflected that:

There is an ever-recurring cycle of three generations. One finds the god, the second arches the narrow temple over him and in doing so fetters him, while the third slides into poverty and takes stone after stone from the sanctuary in order to build meagre and makeshift huts. And then comes one which must seek the god again ...[1]

By and large, many of the organisations and groupings with which we work will be peopled by incumbents of the second and third generations; the creativity of impulse gives way to an institutional imperative which leads ultimately to the carcass of anarchic originality – echoing corridors marked by an increasing poverty of spirit. As social development practitioners, we must have at least some affinity with the first generation if we are to be of any use at all in the project of social renewal. 'Finding the god' is simply Rilke's metaphor for awakening.

We awaken to three phenomena: to ourselves, to the social processes which surround us, and to our responsibility.

With respect, firstly, to ourselves, the Buddhist writer Stephen Batchelor points out that the heart of the four truths underlying the Buddha's teaching lies in the ability to recognise and let go of our incessant craving for the world to be other than it is.[2] The essence of anguish, on the other hand, is to demand that the world conform to our desires and wishes for how it must be. Such desire, that the world conform to our demands, lies at the heart of continual anguish, and forms the chains which bind, the fetters which keep us attached. It results in distraction, boredom, pain and disappointment. For we can never be satisfied, and we are ever focused on a world which is of our own construction, rather than on that which actually surrounds us. We are both unaware, in a very profound sense, and in pain.

On the other hand: 'To let go is like releasing a snake that you have been clutching in your hand.'[3] Batchelor goes on to say: 'By identifying with a craving ... you tighten the clutch and intensify its resistance. Instead of being a state of mind that you have, it becomes a compulsion that has you.'[4] Such complusion is not the sole preserve of those with neuroses and addictions; it is the state of attachment and confusion with which we often approach the world. Or perhaps, turning this around, we may say that we mostly approach the world as addicts and neurotics, demanding that it conform, ever in pain and anguish that it does not. Such (lack of) consciousness has little to do with awakening, or with being able to see the world as it really is. Awakening to ourselves means letting go of our various attachments, as innocuous as many might seem, or as we would wish them to be. Letting go, the practice of non-attachment, lies at the heart of wakefulness. Roger Harrison, a renowned organisation development practitioner and writer, puts it this way (providing a direct link to a new way of seeing): 'The building of intuitive capacity (which is a kind of guidance) comes from not being attached to the fruits. The greatest power in intuition is when least concerned with the results.'[5]

Being awake to ourselves also means getting in touch with our own inner movements, with the paths and patterns of our own processes of becoming. If we do not understand ourselves we will understand little of the organisations and social situations we serve. If we are not able to reflect on our own processes we will not make sense of the processes of others. If we do not understand our own involvement in the phenomena with which we work, then we will not be able to serve those phenomena adequately or developmentally. As within, so without. If we want to intervene responsively, recognising the underlying patterns which form the heartbeat of the social situations which come towards us, then we must engage in practices of self-reflection, for we experience such patterns most powerfully from inside ourselves. Reflection lies at the heart of a social development practice, for it is the engine of wakefulness.

This is wonderfully articulated by the educational consultant Parker Palmer.[6] Individuals rise to leadership, he says, through their ability to manipulate the world around them, to bend that world to their own ends. We rise to leadership through such success in the outer world, and through being seen to be successful; through charisma and force. Yet the more we focus on outer success, on stamping our authority on outer circumstance, the more we ignore our inner world, the world of the spirit which moves us, the world from which we take our being. In doing so we operate more and more out of an unconscious momentum which is driven by outer success. As we ignore our inner world, so awareness diminishes rather than expands, and we reduce ourselves in direct proportion to that outer success. We lose sight of that which makes us whole – consciousness and meaning.

Thus we become dangerous, for our power grows even as our consciousness dims. And the dangers of certain dynamics become manifest – thus may our shadow rear its head unbidden and unrecognised; thus may the paradoxes of power begin to close us off from the new, the other, the marginal; thus may we ignore the pattern of reversal, and impose ourselves upon the world with an increasingly heavy hand. Because we are focused on the outer, and our own development is ignored, we become increasingly blind and increasingly small, and we visit our projections on to the world around us, which we can no longer really see for what it is. We no longer understand; we are intent only on being understood.

The very factors which propel us into leadership precipitate our downfall, and promote pain and discomfort in those we lead. Because we can no longer see ourselves, we can no longer see the other. Developmental leadership demands that we engage in our own development towards increasing consciousness of self, that we pay attention to our own inner patterns of change and emergence, so that we treat the world without rancour or deceit. Interestingly, that which builds leadership in the eyes of the world – outer success

– is also that which, as we learnt earlier, leads to our dominantly logical and analytic frame of mind. To think holistically, to become receptive, we cannot ignore the movements taking place within our souls. It is here, and not only in the outer world, that truly great leadership is incubated, and responsible social intervention nurtured.

Then, we awake to the social processes which surround us. There are two central aspects which we awake to when we facilitate processes of development. We recognise, in the first instance, that processes of development take their own time; while we may facilitate and guide, we may not impose and control (the word 'may' is used to describe, not to proclaim). Every being is developing along its own trajectory, and such development proceeds according to its own needs with respect to pace and rhythm. As Lievegoed puts it, 'All development occurs slowly.'[7]

This observation, seemingly obvious and even innocuous, in fact takes on radical and controversial dimensions within the world in which we operate today. The forces and powers which surround our work as social development practitioners demand over-hasty time-frames, short-term projects and quick results. In the words of a piece of graffiti: 'They don't want it good, they want it by Wednesday.' This approach, more and more ubiquitous, does not allow for social situations, possibilities, ideals, and visions to *develop*. It believes that society can be engineered according to plans set in advance; it assumes, and promotes, a mechanical society which squeezes out inherent life forces and replaces them with a set of controls. This practice results in a set of conditions which make it very difficult for the individual or the group to breathe, unfold and emerge on its path of becoming. It denies this path, and in so doing it sets up increasingly obstructive internal contradictions which make a mockery of any claim towards sustainability.

The lesson for the social development practitioner is clear: we must learn to live without resolution, without clarity, with continuing ambiguity, with patience and with questions. With a deep confidence that if we only do our work well, without attempting to control, or to turn open-ended facilitation of social process into instruments and procedures and tools for application, then the answers will emerge from out of the forces and energies inherent in the life processes themselves. As T.S. Eliot put it, we should look to the sowing, not to the harvest.[8]

It is well to be aware that we work within a milieu which emphasises only the harvest and regards the sowing as a necessary nuisance. Awakening to social process in this sense entails the emergence into a world, which is passionately at odds with all that we have learned to accept in our growth towards so-called adulthood.

A second instance of awakening to the reality of social process results from the recognition that, however well we may do our work, the dynamic of shadow – raised in Chapter 11 – always applies: Every act of good releases the shadow of the good. We can never ignore or push the shadow aside, or assume that we have circumvented it; we have to accept that it rides side by side with the light on the trail of every organism's journey towards becoming. Whatever we do that is good today will raise difficult issues tomorrow.

The implication of such awareness (awakeness) is to treat the shadow with respect and gentleness. Whenever you move ahead, whenever you facilitate the forward movement of others, there will be people and situations, which remain behind in darkness. These must not be rejected but cared for; and they will form the spur to further development of the entire system at a later stage. Treating the shadow with harshness will drive it into a corner and increase its power immeasurably. As Lievegoed puts it: 'No aggravation, no violence, but gentleness. Developing gentleness towards evil, that is the great task on the path ... Gentleness and love are the forces that save the human soul.'[9] To approach our work with force and power, and yet to work with gentleness and love, with compassion and understanding – such sensibility lies at the heart of our practice.

This is necessary as we awaken to the third phenomenon, to our responsibility. As social development practitioners, our responsibility is to facilitate consciousness, to enlarge consciousness, to (quite literally) enable consciousness to emerge. No real change occurs, no such change is sustained, without it being accompanied and fuelled by consciousness. Change itself often feels like the unattainable pot of gold at rainbow's end, especially for those of us involved with facilitating transformation in large and complex systems like bureaucracies and communities and moribund social groupings. In these situations, every shift, increase, enlargement of consciousness, every iota of awareness and understanding of self and other on the part of the social grouping is a move towards a new world. The path of becoming is served, polarities are held in creative tension rather than distorted into bias and prejudice, the world around is permitted to enter us for what it is and not contorted into what we would have it be. The group's life – and thereby the life of other groups as well – becomes that much more interesting to itself. This is, in James Hollis' words, an 'immense gift, a stupendous contribution'.[10]

The psychotherapist Robert Johnson wrote: 'The big question, the question that should confront every modern person is: ... How can we participate in this evolution of consciousness? If we can understand this, ... we can contribute something meaningful. This task is so important because archetypal patterns ... do not necessarily go in a positive direction.'[11] Precisely, then, at this stage in our history, when archetypal patterns and edgeless freedoms are balanced so precariously one against the other that the dangers of delusion and manipulation grow in proportion to the dreams of liberation, the work of facilitating consciousness becomes the hero's task for the modern (and post-modern) person. Johnson goes on to write: 'People used to ask Dr Jung, "Do you think we will make it? Will civilisation survive?" He invariably answered, "If enough people will be conscious."'[12]

It is exactly at this point, when we recognise our responsibility as work in the service of consciousness, that three dominant strands informing this book come together – the psychological work of Jung, the phenomenological work of Goethe, and the transitional poetry of Rilke. All three contribute a similar perspective (though from entirely different points of departure) towards the underpinning of the new discipline, art and practice of social development.

For Jung, the purpose of human life is the creation of consciousness. This takes place through the process of individuation, of becoming, as consciousness is gradually and painfully drawn out of the transformation of the unconscious, which is bequeathed to us as given, as birthright. For Jung, the unconscious and God are one and the same, and the task of transforming the unconscious into consciousness encompasses nothing less than the transformation of God. It is we humans who must contribute towards the unfolding of the world into something new which can never have been before. And thus too does God develop through us.

Consciousness is brought about through encountering and enduring the union of opposites. While such opposites are initially experienced as painful and paralysing conflicts, in fact 'whenever one is experiencing the conflict between contrary attitudes or when a personal desire or idea is being contested by an "other", either from inside or outside, the possibility of creating a new increment of consciousness exists'.[13] Such consciousness, such resolution of opposites, can only take place within the human soul. Jung wrote that, 'Man is the mirror which God holds up before him, or the sense organ with which he apprehends his being.'[14] The Jungian writer Edward Edinger also notes that consciousness, in its resolution of opposites, is the essential carrier of unity in social life, and that it is through consciousness alone that the resolution of conflict in social life may be attained. He attributes to Jung the thought that, 'When enough individuals are carriers of the "consciousness of wholeness", the world itself will become whole.'[15]

Goethe, in developing a new way of observing nature itself, provides us with methodologies and concepts which enable us to become conscious of wholes (and of wholeness and its opposite) in such a way that we can actually begin to see and experience the invisible and intangible meanings which form through the meeting between self and other. In this way, every time we perceive such meaning we contribute not only to wholeness but also to the ongoing evolution and enlargement of our world. Thus for Goethe as well, we transform our world even as we understand it, and the more conscious and aware we become, the more there is of the world in which we live. For Goethe too we are co-creators. And we are also the sense organ of the world, through which nature may begin to know itself, and without which everything remains unconscious and undeveloped.

Goethe provides our first real access to wholeness, to the possibility of seeing intangible wholes in such a way that they may reveal the formative forces which shape the material world, and the archetypal patterns which provide the energy by means of which complex living systems evolve. While Jung provides insight into psychological mechanisms as a means of furthering consciousness in individuals, Goethe provides a phenomenological method – not least through the use of metaphor and imagination – which takes us beyond reductionism and enables us to come to grips with the forces, both psychological and spiritual, which lie within the living system as a complex and composite organism.

Rilke's particular use of poetic metaphor and imagination is so valuable to us as practitioners because it enables the reader to develop a real sensibility for transience, for ambiguity and uncertainty, for the profound greatness and tragedy of the human condition as it attempts to evolve. Rilke never tries to remove the pain and pathos, yet his work abounds with numinosity and with inspiration. Most of all, it penetrates the human condition in a way which never defines or captures or reduces, but always enlarges. On reading it one is always left wondering; there is movement and meta-morphosis always, colours which shift as the day progresses, landscapes which transmute as if through eons of time. And always there is doubt, and always there is certainty.

I mention these things here, as this book draws to a close, because we are always in search of the shortcut, the technique or the tool which we can apply to a situation without thinking. This is the danger which lies at the heart of the analytic way of thinking – we want to reduce, to simplify, to set out an argument in the form of a table, or a listing of bullet points. But the social is not amenable to such treatment. There are no real boundaries here, no easy edges. Who can tell where pain stops and largesse begins? Who can tell whether laughter is rueful or joyful? At what point does creativity change into moribund repetition? Who can plot a relationship on a continuum moving from dependence through interdependence, and who is capable of responding appropriately? When is a social being justified in reacting strongly to its surroundings, and when is this simply a projection of its own inadequacies and fears?

All of these questions can only be answered in the moment and in response to a specific situation, and none of these questions, even at such times, can ever be answered definitively. There is ambiguity and uncertainty always; just as there is always movement and change, and the answer itself will change the circumstances which gave rise to it. In the words of a colleague, we are trying to hold infinity.[16] Under such conditions we must indeed beware the tendency to reduce. The challenge is rather to become discerning practitioners.

As we work with the social in this way, we do far more than merely solve problems. We enable meaning and spirit and original impulse to infuse matter and manifestation. We co-create a living universe within which we are able to develop ourselves towards ever larger artistic horizons. We work with the possible, not only with the given. Every social practitioner who works in this way is not working simply with one particular social grouping but is sowing a seed in the fertile earth of humanity's very soul.

Often, such sowing may seem small and insignificant, even invisible. Yet now that we are witness to the power of the invisible, we may begin to appreciate the value of such work.

Notes

Preface

1. Henri Bortoft, *The Wholeness of Nature* (New York: Lindisfarne Press, 1996), p.11.
2. Donald A. Schon, *The Reflective Practitioner: How Professionals Think in Action* (Aldershot: Academic Publishing Group, 1991).
3. Example taken from David Harding, 'Capacity Building – A Deep Challenge Needs a More Effective Response' (unpublished manuscript, 2000).
4. Willemien le Roux, *Shadowbird* (Cape Town: Kwela Books, 2000), p.25.
5. Vaclav Havel, Speech to World Economic Forum, Geneva 1992.

Chapter 1

1. Rudolf Steiner, *A Theory of Knowledge* (Spring Valley, NY: Anthroposophic Press, 1978), p.7.
2. David Hume, A *Treatise of Human Nature* (New York: Clarendon Press, 1985).
3. As quoted in Ernst Lehrs, *Man or Matter* (London: Faber and Faber Ltd, 1958), p.82.
4. Professor Carel, in his book *Man the Unknown*, as quoted in Lehrs, *Man or Matter*, p.73.
5. Johann Wolfgang von Goethe, *Faust* (Volume One), translated by Phillip Wayne, (London: Penguin Books, 1959), p.18.
6. For example, *Newsweek Magazine*, in a story concerning the search for life on Mars, notes blithely that the discovery of life on Mars would 'shed light on how life emerged from non-life on Earth'; there is simply no question as to which came first. *Newsweek*, 6 December 1999, p.57.
7. See for example, Margaret Wheatley, *Leadership and the New Science: Learning about Organisation from an Orderly Universe* (San Francisco: Berrett-Koehler Publishers, 1992) and Fritjof Capra, *The Web of Life: A New Synthesis of Mind and Matter* (London: HarperCollins Publishers, 1996).
8. Michael Talbot, *Voice*, 22 September 1987.
9. Wheatley, *Leadership and the New Science*.

Chapter 2

1. For the manner in which the argument is presented in this chapter, I am indebted to Arnold Freeman, *Self-Observation* (London: Anthroposophical Publishing Company, 1956).
2. Ibid., p.20.

Chapter 3

1. For the argument which follows, I am indebted to Henri Bortoft, *The Wholeness of Nature* (New York: Lindisfarne Press, 1996).
2. E.E. Pfeiffer, *Life's Resources* (New York: Anthroposophic Press, 1963), p.9.
3. Ibid.
4. Theodor Schwenk, *Sensitive Chaos*, translated by Olive Wicher and Johanna Wrigley (London: Rudolf Steiner Press, 1965), p.13.
5. S.H. Skaife, *African Insect Life* (South Africa: Struik Publishers, 1992), p.51.
6. E.E. Pfeiffer, *Life's Resources*, p.17.
7. Michael Talbot, *Voice*, 22 September 1987.
8. As quoted in Rudolf Haushka, *The Nature of Substance* (London: Vincent Stuart Limited, 1966), p.14.
9. E.E. Pfeiffer, *Life's Resources*, p.5.
10. John Heider, *The Tao of Leadership* (New York: Bantam Books, 1988), p.21.

Chapter 4

1. For the manner in which the argument is presented in this chapter, I am indebted to Henri Bortoft, *The Wholeness of Nature* (New York: Lindisfarne Press, 1996).
2. Ibid., p.301.
3. Ibid., p.304.
4. Ibid., p.309.

Chapter 5

1. As quoted in John Alexander, *Mephistopheles' Anvil* (New York: Rose Harmoney Publications, 1996), p.34.
2. J.D. Salinger, *Raise High the Roof Beam, Carpenters* (London: Penguin Books, 1964), p.9.
3. Valerie Jacobs, *Black and White Shaded Drawing* (London: Rudolf Steiner Press, 1975), p.56.
4. As quoted in Rudolf Haushka, *The Nature of Substance* (London: Vincent Stuart Limited, 1966), p.7.
5. See, for example, Betty Edwards, *Drawing on the Right Side of the Brain* (London: Fontana, 1982); Betty Edwards, *Drawing on the Artist Within* (London: HarperCollins Publishers, 1995); Margaret Colquhoun and Axel Ewald, *New Eyes for Plants* (Stroud: Hawthorn Press, 1996); Frederick Franck, *The Zen of Seeing* (New York: Vintage Books, 1973); Roberta Weir, *Leonardo's Ink Bottle* (Berkeley: Celestrial Arts, 1998).
6. For the original elaboration of this exercise, see Edwards, *Drawing on the Right Side of the Brain*, pp.50–9.
7. Ibid., pp.82–93.
8. Kees Locher and Jos van der Brug, *Workways – Seven Stars To Steer By* (Stroud: Hawthorn Press, 1997), pp.58–9.
9. See for example, Edwards, *Drawing on the Artist Within*, pp.66–95.

Chapter 6

1. I Ching or Book of Changes, translated by Richard Wilhelm (London: Arkana-Penguin Books, 1989), p.282.
2. Johann Wolfgang von Goethe's essay 'Formation and Transformation', as quoted in Henri Bortoft, *The Wholeness of Nature* (New York: Lindisfarne Press, 1996), p.285.
3. Ibid.
4. Marcus Aurelius, *Meditations*, translated by Maxwell Staniforth (London: Penguin Books, 1964), p.72.
5. Goethe, 'Formation and Transformation', in Bortoft, *The Wholeness of Nature*, p.285.
6. Original drawing by Lindy Solomons Cape Town, 2001.
7. As quoted by Blanaid McKinney in his short story *The Outfielder, the Indian Giver*, included in *The New Picador Book of Contemporary Irish Fiction*, edited by Dermot Bolger (Oxford: Picador, 2000), p.584.
8. Taken from one of Rudolf Steiner's meditations, rendered as:

> The stars spake once to man,
> But they are silent now
> To perceive their silence
> Can be a grief for earthly man
> Yet in the silence ripens now
> What human beings say to the stars
> To engage with this speech
> Strengthens our emerging destiny.

9. George Steiner, 'Through that Glass Darkly', in a collection of essays entitled *No Passion Spent* (London: Faber and Faber, 1996), p.347.
10. W.B. Yeats, 'The Circus Animal's Desertion', in *Collected Poems of W.B. Yeats* (London: Macmillan, 1973), p.392.
11. See for example Arthur J. Deikman, 'Bimodal Consciousness', in *The Nature of Human Consciousness*, edited by Robert E. Ornstein (San Francisco: W.H. Freeman, 1973).
12. Nicanor Perlas, *Shaping Globalisation: Civil Society, Cultural Power and Threefolding* (Quezon City: Centre for Development Alternatives, 1999).
13. Ernst Lehrs, *Man or Matter* (London: Faber and Faber Ltd, 1958), p.57.
14. Ibid.
15. James Hollis, *Tracking the Gods* (Toronto: Inner City Books, 1995), p.26.
16. Ibid., p.29.
17. Christopher Fry, *A Sleep of Prisoners*, in *Selected Plays* (Oxford: Oxford University Press, 1992), p.253.

Chapter 7

1. I Ching or Book of Changes, translated by Richard Welhelm (London: Arkana-Penguin Books, 1989), p.283.
2. Joseph Campbell, *The Power of Myth* (New York and London: Doubleday, 1988), p.5.

3. Rainer Maria Rilke, *The Duino Elegies* (New York: W.W. Norton and Company), translated by David Young, p.39.
4. Ibid.
5. Ibid., p.43.
6. Ibid.
7. W.B. Yeats, 'The Second Coming', in *Collected Poems* of W.B. Yeats (London: Macmillan, 1973), p.210.
8. Rainer Maria Rilke, 'Just as the Winged Energy of Delight', in *Rag and Bone Shop of the Heart*, translated by Robert Bly, James Hillman and Michael Meade (New York: Harper Perennial, 1993), p.236.
9. I am indebted to Gavin Andersson for this tale.
10. See, for example, James Hollis, *Tracking the Gods* (Toronto: Inner City Books, 1995).
11. Fritjof Capra, *The Web of Life – A New Synthesis of Mind and Matter* (London: HarperCollins Publishers, 1996), p.154.
12. This exercise is derived mainly from Rudolf Steiner, *Knowledge of the Higher Worlds* (London: Rudolf Steiner Press, 1973), pp.31–44; it is also elaborated in Jorgen Smit, *How to Transform Thinking, Feeling and Willing* (Stroud: Hawthorn Press, 1988), pp.31–3.

Chapter 8

1. I Ching or Book of Changes, translated by Richard Wilhelm (London: Arkana-Penguin Books, 1989), p.298.
2. Ibid., p.282.
3. To follow such experiments in sufficient detail to repeat them, and indeed for an in-depth rendition of the very truncated argument presented here, see Henri Bortoft, *The Wholeness of Nature* (New York: Lindisfarne Press, 1996), pp.36–50.
4. John Heider, *The Tao of Leadership* (New York: Bantam Books, 1988), p.3.
5. William Blake, *Complete Writings* (London: Oxford University Press, 1972), p.149.
6. See Paul Matthews, *Sing Me the Creation* (Stroud: Hawthorn Press, 1966), p.89.
7. Lindsay Clarke, *The Chymical Wedding* (London: Picador, 1989), p.415.
8. Some books, amongst many that may be helpful here, include Rob Nairn, *Diamond Mind* (Cape Town: Kairon Press, 1998); Stephen Batchelor, *Buddhism Without Beliefs* (London: Bloomsbury, 1997); David Fontana, *The Mediator's Handbook* (Shaftesbury: Element Books, 1998); Jorgen Smit, *How to Transform Thinking, Feeling and Willing* (Stroud: Hawthorn Press, 1989); Rudolf Steiner, *Knowledge of the Higher Worlds* (London: Rudolf Steiner Press, 1973).
9. Rudolf Steiner, *Practical Training in Thought* (New York: Anthroposophic Press, 1977), p.6.
10. Bortoft, *The Wholeness of Nature*, p.247.

Chapter 9

1. Ernst Lehrs, *Man or Matter* (London: Faber and Faber Ltd, 1958), p.112.
2. A point impressed upon me by Mario van Boesschoten many years ago.

3. This exercise is partly derived from Jorgen Smit, *How to Transform Thinking, Feeling and Willing* (Stroud: Hawthorn Press, 1998), pp.16–19.

Chapter 10

1. In a composite social organism it will not happen simultaneously throughout. Different sections, departments, teams may move at different paces, though each will always be affected by the others.

Chapter 11

1. Robert A. Johnson, *Owning Your Own Shadow* (San Francisco: Harper, 1991), p.13.
2. Sue Soal, a fellow practitioner from the Community Development Resource Association.
3. I am obliged to Mario van Boeschoten for introducing me to the exercise.

Chapter 12

1. For further exploration in this regard, see Allan Kaplan, *The Development Capacity* (Geneva: United Nations Non-governmental Liaison Service, 1999), particularly chapter 2.
2. Verbal communication during a training workshop conducted in 1994.
3. Bernard Lievegoed, *Towards the Twenty-First Century: Doing the Good* (Bristol: Steiner Press, 1979), p.38.
4. This exercise was introduced to me by Fritz Glasl.

Chapter 13

1. Please refer back to Chapter 4 if necessary.
2. See, for example, Bernard Lievegoed, *The Developing Organisation* (London: Tavistock Publications, 1973), amongst others.
3. See Max de Preez, *Leadership is an Art* (New York and London: Doubleday, 1991), pp.97–101.
4. Vaclav Havel, *Disturbing The Peace*, translated by Paul Wilson (New York: Vintage Books, 1991), p.11.
5. Fritjof Capra, *The Web of Life: A New Synthesis of Mind and Matter* (London: HarperCollins Publishers, 1996), pp.163–4.
6. Lao Tzu, Tao Te Ching, translated by Aleister Crowley (York Beach, ME: Samuel Weiser Inc., 1995), p.26.
7. Ibid.
8. Peter Senge et al., *The Dance of Change: The Challenges of Sustaining Momentum in Learning Organisations* (London: Nicholas Brealey Publishing, 1999), p.157.
9. Ibid.
10. Quoted in 'De -Profundis' by Oscar Wilde, in *The Works of Oscar Wilde* (London: Spring, 1977).

Chapter 14

1. Olive Schreiner, *The Story of an African Farm* (Jeppestown: A.D. Donker Pty Ltd, 1998).
2. See, for example, Rudolf Steiner, *Knowledge of the Higher Worlds* (London: Rudolf Steiner Press, 1973), pp.19–31.

Chapter 15

1. As quoted by Chris Roche 'Impact Assessment: Seeing the Wood and the Trees', in *Development Practice*, volume 10, edited by Deborah Eade (Oxford: Oxfam, 2000), p.543.
2. Michael Brater, Ute Buchele and Hans Herzer, *Eurythmy in the Workplace* (Chicago: Rudolf Steiner Books, 1998), p.2.
3. Meas Nee (as told to Joan Healy), *Towards Restoring Life in Cambodian Villages* (Phnom Penh: JSRC, 1999), p.40.
4. Julia Cameron, *The Artist's Way* (New York: G.P. Putnam's Sons, 1992), p.66.
5. Robert Johnson, *Inner Work* (San Francisco: Harper Books, 1989), p.23.
6. See Paul Matthews, *Sing Me the Creation* (Stroud: Hawthorn Press, 1996), pp.40–1.

Chapter 16

1. This quote is attributed to St Francis of Assisi.
2. I am indebted to my old colleague Mzwandile Msoki.
3. D.H. Lawrence, 'Song of a Man who has Come Through', in *The Rag and Bone Shop of the Heart*, edited by Robert Bly, James Hillman, Michael Meade (New York: Harper Perennial, 1992), p.20.
4. As quoted in the Annual Report of the Community Development Resource Association, 1997/98 *Crossroads: A Development Reading* (Cape Town: CDRA).
5. Elizabeth Kubler-Ross, *On Death and Dying* (London: Routledge, 1973).

Chapter 17

1. Bernard Lievegoed, *Forming Curative Communities*, translated by Stephen Briault (London: Rudolf Steiner Press, 1978), pp.9–14.
2. Ibid., p.9.
3. Ibid., p.12.
4. Mario van Boeschoten, personal communication.

Chapter 18

1. Rumi, *The Essential Rumi*, translated by Coleman Barks (San Francisco: Harper, 1995), p.178.
2. An interesting case study which describes some of this particularly succinctly and well can be found in Michael Brater, Ute Buchele and Hans Herzer, *Eurythmy in the Workplace* (Chicago: Rudolf Steiner Books, 1998), particularly pp.6–8.

Chapter 19

1. Peter Senge et al., *The Dance of Change: The Challenges of Sustaining Momentum in Learning Organisations* (London: Nicholas Brealey Publishing, 1999), p.157.
2. Jacqueline Leslie Roberts, *Wilderness, a Circle of Courage, and the Wisdom of Elders* (Cape Town: Educo Africa, 1999), p.6. Copyright, *The Transformation of the Child and Youth Care System*.
3. Jeremy Naydler, *Goethe on Science: Selected Quotations*, (Edinburgh: Floris Books, 1996), p.66.
4. Ibid., p.67.
5. Ibid., p.60.
6. Ibid., p.29.

Chapter 20

1. Fritz Glasl, *The Enterprise of the Future* (Stroud: Hawthorn Press, 1997), p.65.
2. Rudolf Steiner, *Practical Training in Thought* (New York: Anthroposophic Press, 1977), p.6.
3. Ibid.
4. Henri Bortoft, *The Wholeness of Nature* (New York: Lindisfarne Press, 1996), p.247.
5. Christopher Alexander, *The Timeless Way of Building* (New York: Oxford University Press, 1979), p.13.
6. Ibid.
7. Eugen Herrigel, *Zen in the Art of Archery* (New York: Vintage Books, 1989), p.59.
8. This exercise was initially derived from a reading of Alexander, *The Timeless Way of Building*, p.12.
9. Ibid., p.91.

Chapter 21

1. Henri Bortoft, *The Wholeness of Nature* (New York: Lindisfarne Press, 1996), p.245.
2. Rainer Maria Rilke, *Letters to a Young Poet* (Novoto, CA: New World Library, 2000), pp.55–6.
3. I am obliged to Carol Liknaitsky for this reading of the Grail legend.
4. Christopher Alexander, *The Timeless Way of Building* (New York: Oxford University Press, 1979), p.17.
5. Ibid., p.25.
6. Ibid., p.26.
7. Ibid., p.162.
8. Rilke, *Letters to a Young Poet*, p.35.

Chapter 22

1. Lao Tzu, Tao Te Ching, translated by Aleister Crowley (York Beach, ME: Samuel Weiser, 1995), p.25.
2. Ibid.

3. Ibid., p.18.
4. I am indebted to Mario van Boeschoten for this insight.
5. Rumi, *The Essential Rumi*, translated by Coleman Barks (San Francisco: Harper, 1995), p.28.
6. Eugen Herrigel, *Zen in the Art of Archery* (New York: Vintage Books, 1989), p.37.
7. Douglas Steere, *Conjectures of a Guilty Bystander* (New York: Dell, 1968), p. 86.
8. As quoted by Blanaid McKinney, in his short story 'The Outfielder, the Indian Giver', included in *The New Picador Book of Contemporary Irish Fiction*, edited by Dermot Bolger (Oxford: Picador, 2000), p.601.
9. Esther de Waal, *Living with Contradiction* (London: Fount Paperbacks, 1989), p.118.
10. Ibid., p.43.
11. As quoted in ibid., p.89.
12. In different format aspects of this exercise can be found in Jorgen Smit, *How to Transform Thinking, Feeling and Willing* (Stroud: Hawthorn Press, 1998), p.36.

Chapter 23

1. Rainer Maria Rilke, *Diaries of a Young Poet*, translated by Edward Snow and Michael Winkler (New York: W.W. Norton and Company, 1997), p.15.
2. Stephen Batchelor, *Buddhism Without Beliefs* (London: Bloomsbury, 1998), pp.3–13.
3. Ibid., p.9.
4. Ibid.
5. Roger Harrison, *A Consultant's Journey: A Professional and Personal Odyssey* (New York: McGraw-Hill, 1995), p.181.
6. Parker J. Palmer, 'Leading From Within', speech published by the Indiana Office for Campus Ministries in Indianapolis, with support from the Lilly Endowment, 1990.
7. Bernard Lievegoed, *Battle for the Soul* (Stroud: Hawthorn Press, 1994), p.73.
8. T.S. Eliot, 'Choruses from "The Rock"', in *The Complete Poems and Plays of T.S. Eliot* (London: Faber and Faber, 1969), p.148.
9. Lievegoed, *Battle for the Soul*, p.88.
10. James Hollis, *Creating a Life* (Toronto: Inner City Books, 2001), p.19.
11. Robert Johnson, *Balancing Heaven and Earth* (New York: Harper, 1998), p.133.
12. Ibid.
13. Edward Edinger, *The Creation of Consciousness* (Toronto: Inner City Books, 1984), p.18.
14. C.G. Jung, *Letter* (28 March 1953). Cited by Aniela Jaffe in *Phases in Jung's Life*, Spring 1972, p.136.
15. Edinger, *The Creation of Consciousness*, p.32.
16. I am indebted to Sue Soal for alerting me to William Blake's riveting phrase.

Index

Compiled by Sue Carlton

Printed and bound by CPI Group (UK) Ltd, Croydon, CR0 4YY

27/10/2024

14580225-0003